A GUIDE TO WRITING
AS AN ENGINEER

A GUIDE TO WRITING AS AN ENGINEER

SECOND EDITION

David Beer

Department of Electrical and Computer Engineering
University of Texas at Austin

David McMurrey

Austin Community College

JOHN WILEY & SONS, INC.

Acquisitions Editor   Joseph Hayton
Marketing Manager   Jennifer Powers
Production Editor   Sandra Dumas
Senior Designer   Karin Kincheloe
Production Management Services   Argosy Publishing
Cover art   @Harvey Chan/i2i Art

This book was typeset in 10/12 Times Roman by Argosy and printed and bound by Courier (Westford). The cover was printed by Phoenix Color Corp.

The paper in this book was manufactured by a mill whose forest management programs include sustained yield harvesting of its timberlands. Sustained yield harvesting principles ensure that the number of trees cut each year does not exceed the amount of new growth.

This book is printed on acid-free paper. ∞

Beer, David; McMurrey, David
A Guide to Writing As an Engineer—Second Edition.

ISBN 0-471-43074-9

Printed in the United States of America.

10 9 8 7 6 5 4 3 2

PREFACE

A Guide to Writing as an Engineer, Second Edition, is intended for professional engineers, engineering students, and students in all other technical disciplines. It not only addresses important writing concepts that apply to professional engineering communication, but also deals with the content, organization, format, and style of specific kinds of technical writing such as reports, business letters, office memoranda, and email. The book also covers oral presentations and how to find engineering information, both in traditional ways and on the Internet. The final chapter is concerned with questions of ethics and technical writing, and also provides a citation system for ensuring that all written work and graphics are thoroughly documented when necessary.

WHAT'S NEW IN THIS EDITION

Besides adding a new chapter on ethics and documentation, we have also provided a separate chapter on how to format and use graphics and tables in written reports. Our chapter on accessing engineering information has been completely updated, both for Web and library research, and many new and current URLs have been provided. We have also updated and revised other chapters and have often reorganized with an eye to easier accessibility. Relevant quotes from industrial and academic authorities have been increased, and exercises at the end of each chapter have in some cases been improved.

WHO THIS BOOK IS FOR

The idea for this book grew from our experience in industry and the engineering classroom, and also from our wish to write a text that is practical rather than theoretical, and that devotes all its pages to the writing concerns of working engineers and those planning to become engineers. A common complaint among engineers and engineering students is that there is no helpful book on writing aimed specifically at them. Most technical writing texts focus, as their titles imply, on the entire field of technical writing. In other words, they aim to prepare readers and students for a complete knowledge of everything a technical writer is called on to do: They train people to become technical writers.

Engineers need to know how to write as much as anyone, but few have time to become technical writers. They are required to write all kinds of short documents and help in writing a variety of longer ones, but few need to acquire the skills of an advanced copy editor, graphic artist, or publisher. For most, engineering is their focus, and although advancement to management might bring increasing communication-related responsibilities and opportunities, these will for the most part still be focused on engineering and closely related disciplines.

Thus we have written this book so that engineers and engineering students will have a resource that stays close to the real concerns they have in their everyday professional life. These are concerns we have identified over our combined forty years of teaching and working in industry. Our perspective is the reason we give short shrift to some topics a technical writing book might spend several pages on, yet devote a chapter or two to what a traditional text might relegate to an appendix. These choices and priorities reflect what we have found to be important to the audience of this book—engineers and students of technical disciplines.

Our book is also written with the classroom in mind. It can serve as a text in a writing course for science and engineering majors, or indeed for any student who wants to become familiar with writing in the technology professions. Teachers will find the exercises at the end of each chapter good starting points for discussion and homework. Others who use the book will find these exercises well worth thinking about because they are designed to open up the material in the chapters to a larger context than the individual's own experience. The chapters themselves can be read from beginning to end, of course, but readers can also rely on the table of contents and index, as well as headings and subheadings within chapters, to get them where they need to go. Thus the book can function not only as a textbook but also as a reference for industrial writing and research, oral presentations, documentation of research, and ethical practice in technical communication.

WHAT IS IN THIS BOOK

To keep our book focused squarely on the concerns of engineers, engineering students, and students of technical disciplines, we have organized the chapters in the following way.

Chapter 1, "Engineers and Writing," describes the importance of writing in your professional engineering life and provides a conceptual framework for understanding what impedes the communication process.

Chapter 2, "Some Guidelines for Good Engineering Writing," reviews a dozen essential requirements and guidelines for producing effective engineering documents.

Chapter 3, "Eliminating Intermittent Noise in Writing," reviews specific writing problems that can cause communication problems in engineering writing.

Chapter 4, "Writing Letters, Memoranda, and Email," moves from the conceptual foundations covered in the preceding chapters to one of the most important applications of writing: professional correspondence. This chapter covers format and style for office memoranda, business letters, and email. The chapter has a special section devoted to professional communications on the Internet.

Chapter 5, "Writing Common Engineering Documents," provides content, format, and style recommendations for such common engineering documents as inspection and trip reports, laboratory reports, specifications, progress reports, proposals, instructions, and recommendation reports.

Chapter 6, "Writing an Engineering Report," provides a standard format for the engineering report, with special emphasis on content and style for components such as the cover, transmittal letter, title page, table of contents, executive summary, graphics, tables, and documentation.

Chapter 7, "Constructing Tables and Graphics," focuses in detail on guidelines and techniques for incorporating illustrations and tables into technical documents.

Chapter 8, "Accessing Engineering Information," outlines strategies you can use to find information in traditional libraries as well as on their contemporary online counterparts. This chapter contains a special section on finding information on the World Wide Web and using resources available on the Internet.

Chapter 9, "Engineering Your Presentations," reviews plans and techniques you can use to prepare and deliver technical presentations, either individually or as part of a team.

Chapter 10, "Writing to Get an Engineering Job," covers the content, organization, style, and format for application letters and resumes—some of the main tools you'll use for getting engineering jobs whether you are a student or an engineer desiring a change.

Chapter 11, "Documentation and Ethics in Engineering Writing," emphasizes the need to avoid plagiarism and to document all research fully and reliably. Tools to do this are provided in the chapter, as are sample citations for various kinds of written documents and other sources. Other ethical pitfalls that a writer may encounter are also considered, and two codes of ethics are provided to enable an engineer to substantiate his or her ethical position.

ACKNOWLEDGMENTS

Many talented people have played a part, directly or indirectly, in bringing this book to print. We recognize the input of recent and past students in the Department of Electrical and Computer Engineering (ECE) at the University of Texas at Austin who are now successfully in industry or graduate school. For this second edition, we particularly appreciate the input of Vicky Cabrera, Darryl Gilbert, Amir Haq, Justin Israelson, Nancy Jazrawi, and Jeffrey Ploetner—all University of Texas ECE seniors—and of a number of engineer friends at Advanced Micro Devices in Austin.

We would like to thank those professors who assisted us in reviewing the manuscript as it was developed. They include: Thomas Ferrara, California State University, Chico; Jay Goldberg, Marquette University; David P. Jackson, McMaster University; Jon A. Leydens, Colorado School of Mines; Jeanne Lindsell, San Jose State University; Scott Mason, University of Arkansas; Geraldine Milano, New Jersey Institute of Technology; Michael P. Polis, Oakland University; Heather Sheardown, McMaster University; and Marie Zener, Arizona State University.

We also are grateful for the help of Teresa Ashley, reference librarian at Austin Community College; Susan Ardis, head librarian, Engineering Library, University of Texas at Austin; Professor W. Mack Grady, Department of Electrical and Computer Engineering, University of Texas at Austin; Gerald Kanapathy of Plumtree Software; and Randy Schrecengost, an Austin-based professional engineer. Moreover, the efficient and excellent work of Cindy Kogut, our copy editor, and Adriana Lavergne, our production editor, must be gratefully recognized—we greatly admire their skills.

And of course we thank our families for their love, understanding, and support while writing this book: Ruth, Natasha, Phoebe, Patrick, and Jane.

Austin, Texas 2004

CONTENTS

11. DOCUMENTATION AND ETHICS IN ENGINEERING WRITING   233

1

ENGINEERS AND WRITING

Communication skills are extremely important. Unfortunately, both written and oral skills are often ignored in engineering schools, so today we have many engineers with excellent ideas and a strong case to make, but they don't know how to make that case. If you can't make the case, no matter how good the science and technology may be, you're not going to see your ideas reach fruition.

> George Heilmeier, corporate executive of Bellcore, in "Educating Tomorrow's Engineers," *ASEE Prism*, May/June 1995

Poor communication skill is the Achilles' heel of many engineers, both young and experienced—and it can even be a career showstopper. In fact, poor communication skills have probably claimed more casualties than corporate downsizing.

> H. T. Roman, "Be a Leader—Mentor Young Engineers," *IEEE USA Today's Engineer*, November 2002

Many engineers and engineering students dislike writing. After all, don't we go into engineering because we want to work with machines, instruments, and numbers rather than words? Didn't we leave writing behind us when we finished English 101? We may have hoped so, but the fact remains—as the chapter-opening quotes so bluntly indicate—that to be a successful engineer you must be able to write (and speak) effectively. Even if you could set up your own lab in a vacuum and avoid communication with all others, what good would your ideas and discoveries be if they never got beyond your own mind?

If you personally feel you haven't mastered writing skills in college, the fault probably is not entirely yours. Few engineering colleges offer adequate (if any) courses in technical writing, and many students find that what writing skills they did possess are badly rusted from lack of use by the time they graduate with an engineering degree. Ironically, most engineering programs devote less than 5% of their curriculum to communication skills—the very skills that many engineers will use some 20% to 40% of their working time. Even this percentage usually increases with promotion, which is why many young engineers eventually find themselves wishing they had taken more writing courses.

But rather than dwell on the negative, let's look at the needs and opportunities that exist in engineering writing, and then see how you can best remove barriers to becoming an efficient and effective writer. You'll soon find that the skills you need to write well are no harder to acquire than many of the technical skills you have already mastered as an engineer or engineering student. First, here are four factors to consider:

1. Engineers write a lot.

2. Engineers write many kinds of documents.

3. A successful engineering career requires strong writing skills.

4. Engineers can learn to write well.

ENGINEERS WRITE A LOT

Many engineers spend over 40% of their work time writing, and usually find the percentage increases as they move up the corporate ladder. It doesn't matter that much of this writing is now sent through electronic mail (email); the need for clear and efficient prose is the same whether it appears on a computer monitor or a sheet of paper. Much written material first read on a screen ends up being printed out on paper anyway—and the likelihood of a completely paperless office, workshop, or engine room still seems pretty remote.

An engineer told us some years ago that while working on the B-1B bomber, he and his colleagues calculated that all the proposals, regulations, manuals, procedures, and memos the project generated weighed almost as much as the bomber itself. Most large ships carry several tons of maintenance and operations manuals. Two trucks were needed to carry the proposals for the ill-fated supercollider project from Texas to Washington. John Naisbitt estimated in his book *Megatrends* over 20 years ago that some 6000 to 7000 scientific articles were being written every day, and even then the amount of recorded scientific and technical information in the world was doubling every five and a half years.

Who generates and transmits—in writing or orally—all this material, together with countless memos, reports, proposals, manuals, and other technical information?

Tuesday's Schedule	
2/25/03	
7:30	Arrive, *read and reply to overnight emails* from overseas.
8:00	Work on project.
10:30	*Meet with* project manager to *write answer to* department head request.
11:00	*Write up a request* to obtain needed technical support.
11:30	Lunch.
12:00	*Meet with* server group about *submitted application* to fix process problems.
12:20	*Reply to* emails from Sales about prospective customers' *technical questions.*
12:30	*Write to* software vendor about how our product works with their plans.
1:00	*Give presentation to* server hosting group *to explain* what my group is doing.
2:00	*Join* the team *to write up* weekly *progress report.*
2:30	*Write emails to* update customers on the status of solving their problems.
2:45	*Write email reply to* questions about *knowledge base article I wrote.*
3:00	*Meet with* group *to discuss project goals* for next four months.
3:30	*Meet with* group *to create presentation of findings* to project management.
4:00	Work on project.
5:00	Leave for day.

Figure 1-1   The working day of a typical engineer calls for plenty of communication skills.

Engineers. Perhaps they get some help from a technical editor if their company employs one, and secretaries may play a part in some cases. Nevertheless, the vast body of technical information available in the world today has its genesis in the writing and speaking of engineers, whether they work alone or in teams. Figure 1-1 shows the response we got when we randomly asked an engineer friend, who works as a software deployment specialist for Plumtree Software, Inc., to outline a typical day at his job (our italics indicate where communication skills are called for).

ENGINEERS WRITE MANY KINDS OF DOCUMENTS

As mentioned earlier, few engineers work in a vacuum. Throughout your career you will interact with a variety of other engineering and nonengineering colleagues, officials, and members of the public. Even if you don't do the actual engineering work, you may have to explain how something was done, should be done, needs to be changed, must be investigated, and so on. The list of all possible engineering situations and contexts in which communication skills are needed is unending. Figure 1-2 identifies just some of the documents you might be involved in producing during your engineering career. (It's worth noting that not all companies label reports by the same name or put them in the same categories as we have. Also, many of these reports would obviously overlap into more than one of the "files" we have somewhat arbitrarily placed them in.)

As the 21st century gets under way, electronic communication is rapidly replacing much hard copy. Used for anything from quick, pithy notes and memos to complete multivolume documents, email is becoming the most popular form of written communication. Yet this does not in any way change the need for clarity and organization in engineering writing, and whatever the future holds, solid skills in clear and efficient writing, and the ability to adapt to many different document specifications, will probably be necessary for as long as humans communicate with each other.

A SUCCESSFUL ENGINEERING CAREER REQUIRES STRONG WRITING SKILLS

In the engineering field, you are rarely judged solely by the quality of your technical knowledge or work. People also form opinions of you by what you say and write.

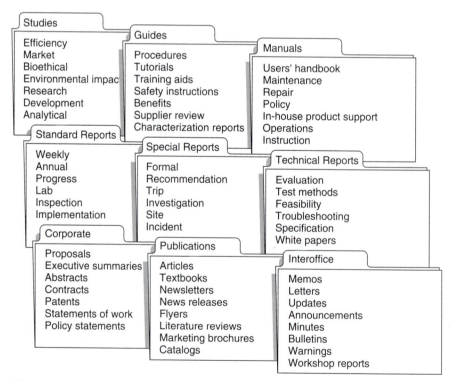

Figure 1-2   Throughout their careers, engineers write many kinds of documents in various contexts and with different purposes and audiences.

When you write a memo or report, talk to members of a group, deal with vendors on the phone, or attend meetings, the image others get of you is largely formed by how well you communicate. Even if you work for a large company and don't see a lot of high-level managers, they can still gain an impression of you by the quality of your written reports as well as by what your immediate supervisor tells them.

Writing on "What Courses Should I Take?" in a newsletter for engineering students at the University of Texas at Austin, Richard C. Levine, then manager of hardware planning at Bell Northern Research, stated:

> *Another fundamental is the ability to read with comprehension and to write clearly and correctly. . . . I can't emphasize enough that both of these skills are extremely important. I am not a picky person when it comes to spelling and grammar, but when I see a report or memo that has repeated errors I immediately question the ability and dedication of the person who wrote it. Why didn't they take the time and effort to do it right? Most of the successful engineers I know write clear, well-organized memos and reports. Engineers who can't write well are definitely held back from career advancement.*
>
> *The Current* 3, no. 1 (April 1987), p. 1.

Opinions like this are common among engineering management. The majority of people who have advanced in an engineering career will tell you the same thing. If you don't believe us, ask them!

Two relatively recent trends now making communication skills even more vital to the engineering profession are *specialization* and *accountability*. Because of the advancement and specialization of technology, engineers are finding it increasingly difficult to communicate with one another. Almost daily, engineering fields once considered unified become progressively fragmented, and it's quite possible for two engineers with similar degrees to have large knowledge gaps when it comes to each other's work. In practical terms, this means that a fellow engineer may have only a little more understanding of what you are working on than does a layperson. These gaps in knowledge often have to be bridged, but can't be unless specialists have the skills to communicate clearly and effectively with each other.

In addition to communicating with one another, engineers must be able to communicate with the public, since engineers and their companies are now being held much more accountable by the public. As the director of the Center for Engineering Professionalism at Texas Tech University puts it,

> *The expansiveness of technology is such that now, more than ever, society is holding engineering professionals accountable for decisions that affect a full range of daily life activities. Engineers are now responsible for saying: "Can we do it, should we do it, if we do it, can we control it, and are we willing to be accountable for it?" There have been too many "headline type" instances of technology gone astray for it to be otherwise. . . . Pinto automobiles that burn*

when hit from the rear, DC-10s that crash when cargo doors don't hold, bridges that collapse, Hyatt Regency walkways that fall, space shuttles that explode on national TV, gas leaks that kill thousands, nuclear plant accidents, computer viruses, oil tanker spills, and on and on.

> Engineering Ethics Module, Murdough Center for Engineering Professionalism, Texas Tech University, Lubbock, Texas. http://www. murdough.ttu.edu/EthicsModule/Ethics Module.htm. Accessed February 5, 2003.

Sadly, this statement would now have to include the tragic disaster of the *Columbia* space shuttle, as we all know.

People do want to know *why* a space shuttle crashed (after all, their taxes paid for the mission). They want to know if it really is safe to live near a nuclear reactor or high-power lines. The public—often through the press—wants to know if a plant is environmentally sound or if a project is likely to be worth the tax dollars. Moreover, there is no shortage of lawyers ready to hold engineering firms and projects accountable for their actions. All this means that engineers are being called upon to explain themselves in numerous ways and must now communicate with an increasing variety of people—many of whom are not engineers.

ENGINEERS CAN LEARN TO WRITE WELL

Here are the words of Norman Augustine, chairman and CEO of Martin Marietta Corporation and also chair of the National Academy of Engineering:

> *Living in a "sound bite" world, engineers must learn to communicate effectively. In my judgment, this remains the greatest shortcoming of most engineers today—particularly insofar as written communication is concerned. It is not sensible to continue to place our candle under a bushel as we too often have in the past. If we put our trust solely in the primacy of logic and technical skills, we will lose the contest for the public's attention—and in the end, both the public and the engineer will be the loser.*

> Norman R. Augustine, in *The Bridge*, The National Academy of Engineering, vol. 24, no. 3 (Fall 1994), p. 13.

Writing is not easy for most of us, and like programming, painting, or playing the bagpipes, good writing takes practice. A lot of truth lies in the adage that no one can be a good writer—only a good *re*writer. If you look at the early drafts of the most famous authors' works you will see various scribbling, additions, deletions, rewordings, and corrections where they have edited their text. So don't expect to produce a masterpiece of writing on your first try. Every initial draft of a document, whether it's a 1-page memo or a 50-page set of procedures, needs to be worked on and improved before being sent to its readers.

As an engineer, you have been trained to think logically. In the laboratory or workshop, you are concerned with precision and accuracy. From elementary and secondary school you already possess the skills needed for basic written communication, and every day you can see samples of clear writing in newspapers, weekly news magazines, and popular journal articles. Thus you are already in a good position to become an effective writer partly by emulating what you've already been exposed to. All you need is some instruction and practice. This book will give you plenty of the former, and your engineering career will give you many opportunities for the latter. Meanwhile, keep in mind that as an engineering professional you will frequently have to communicate through a variety of documents and media, you will enhance your career by being able to do so, and you may even find that it can be fun!

NOISE AND THE COMMUNICATION PROCESS

Have you ever been annoyed by interference on your television screen during a favorite program? Perhaps a neighbor was talking on CB radio, and the transmission did nasty things to your reception. Or maybe you couldn't hear a friend clearly on the telephone because someone was using the vacuum cleaner in the next room or the stereo was booming.

In each case, what you were experiencing was noise interfering with the transmission of information. Whenever a message is sent, someone is sending it and someone else is trying to receive it. In communication theory, the sender is the *encoder*, and the receiver is the *decoder*. The message, or *signal*, is sent through a *channel*, usually speech, writing, or some other conventional set of signs. Anything that prevents the signal from flowing clearly through the channel from the encoder to the decoder is *noise*. Figure 1-3 illustrates this concept. Note how all our actions involving communication are "overshadowed" by the possibility of noise.

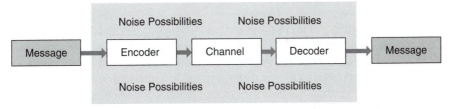

Figure 1-3   In noise-free technical communication, the signal flows from the encoder (writer, speaker) to the decoder (reader, listener) without interruption or ambiguity. When this occurs, the received message is a reliable version of the sent one.

Applying this concept to engineering writing, we can say that anything causing a reader to hesitate in uncertainty, frustration, or even amusement is noise. The following sentences contain just a few samples of written noise:

When they bought the machine they werent aware of it's shortcomings.

They were under the allusion that the project could be completed in six weeks.

There was not a sufficient enough number of samples to validate the data.

Our intention is to implement the verification of the reliability of the system in the near future.

In the first sentence, two apostrophe problems cause noise. A reader might trip over these glitches and momentarily be distracted from the sentence's message (or at least waste time thinking how much smarter he or she is than the writer). The same might be said for the confusion between *allusion* and *illusion* in the second sentence. The third sentence is noisy because of the redundancy and wordiness it contains. Wouldn't you rather just read *There weren't enough samples to validate the data?* The final example is a monument to verbosity. With the noise removed, it simply says: *We want to verify the system's reliability soon.*

It's relatively easy to identify and remove simple noise like this. More challenging is the kind of noise that results from fuzzy and disorganized thinking. Here's a notice posted on a professor's door describing his office hours:

I open most days about 9 or 9:30, occasionally as early as 8, but some days as late as 10 or 10:30. I close about 4 or 4:30, occasionally around 3:30, but sometimes as late as 6 or 6:30. Sometimes in the mornings or afternoons, I'm not here at all, but lately I've been here just about all the time except when I'm somewhere else, but I should be here then, too.

Academic humor, maybe, but it's not hard to find writing in the engineering world that is equally difficult to interpret, as this excerpt from an industrial procedures manual shows:

If containment is not increasing or it is increasing but MG Press is not trending down and PZR level is not decreasing, the Loss of Offsite Power procedure shall be implemented, starting with step 15, unless NAN-S01 and NAN-S02 are de-energized in which case the Reactor Trip procedure shall be performed. But if the containment THRSP is increasing the Excess Steam Demand procedure shall be implemented when MG Press is trending down and the LIOC procedure shall be implemented when the PZR level is decreasing.

The point isn't just that noise in a written document causes anything from momentary confusion to a complete inability to understand a message. Inevitably, noise costs money—or to put it graphically,

Noise = $$$$

According to engineer Bill Brennan, a senior member of the technical staff at Advanced Micro Devices in Austin, Texas, it costs a minimum of $200 to produce one page of an internal technical report and at least five times that much for one page of a technical conference report. As you learn to reduce noise in your writing, you will become an increasingly valuable asset to your company.

Noise can also occur in spoken communication, of course, as you will see in Chapter 9. For now, maybe you can recall how often you've been distracted by a speaker's monotonous tone, nervous cough, clumsy use of notes, or indecipherable graphics—while you just sat there, a captive audience.

The following chapters contain advice, illustrations, and strategies to help you learn to avoid noise in your communication. Try to keep this concept of noise in mind when you write or edit, whether you are working on a 5-sentence memo or a 500-page technical manual. Throughout your school years you may have been reprimanded for "poor writing," " mistakes," "errors to be corrected," "choppy style," and so on, but as an engineer it might be better to think in terms of *noise to be eliminated from the signal.* For efficient and effective communication to take place, the signal-to-noise ratio must be as high as possible. To put it another way, you need to filter as much noise out of your communication as you can.

CONTROLLING THE WRITING SYSTEM

Engineers frequently design, build, and manage systems made up of interconnected parts. Controls have to be built into such systems to guarantee they function correctly and reliably and produce the desired result. The machinery used to mill propeller shafts for large ships must be guided by a control system to ensure that correct tolerances and other specifications are met. If the ATM machine chews up your bank card and spits it back out to you in place of the $100 you had hoped for, you'd claim the system is not working right—or that it is out of control. The system is only functioning reliably if the input (your bank card) produces the desired output (your $100).

What has this got to do with writing? Well, we can view language as a *system* made up of various components such as sounds, words, clauses, sentences, and so on. Whenever we speak or write, we use this system, and like other systems, it must be controlled if it is to do its job right. The person who supposedly wrote in an accident report *Coming home, I drove into the wrong house and collided with a tree I didn't have* was obviously unable to express what really happened. The input (thought) to the system (language) did not have the desired output (meaning) because the writer was not in control of the system, or was not thinking clearly.

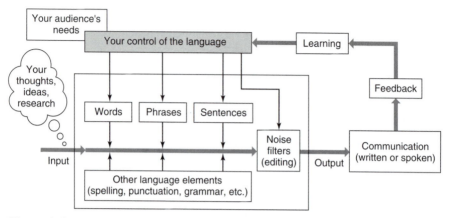

Figure 1-4 The process of communicating can be illustrated as a system with an input and output. How well the input is processed once it is in the system (i.e., how well you convey your information to others) will determine the impact of your message. From the response (feedback) you get, you will learn how to further improve the process.

In the same way, an instruction like *Pour the concrete when it is above 40°F* indicates a lack of language control because the writer is not clearly stating whether the concrete or the weather must meet the specification of "above 40°F." Thus, you might think of language as a system or even a tool you can learn to control so that it will do exactly what you want it to. Learning to control language, namely, to write and speak so you get desired results or feedback, is really not much different from training yourself to operate complex machinery or software systems. With some help and effort, you can train yourself to eliminate most, if not all, noise that might occur when you transfer information by means of writing and speaking. Figure 1-4 depicts how this works. Note how the end product of your communication is often "feedback," which will give you an indication of how well you are using the language system.

If you get the response (or feedback) you want from your communication, you can be pretty sure you have communicated well. A proposal accepted, a part promptly delivered, a repair quickly made, an applied-for promotion awarded—these are just a few examples of the payback from effective communication. To put it another way, if you learn to efficiently control the tool you are using (language) so that it's noise free, you will produce clear and effective written documents that get results.

EXERCISES

1. Ask three professional engineers about the amount and kinds of writing they do on the job. How much of their time is spent writing each day? Is the amount of writing they do

related to how long they have been with their company? In what ways do they feel their writing skills have helped (or hindered) them in their careers so far? Do they get any help with their writing from secretaries, peers, or technical writers? What is the attitude of their superiors toward clear writing?

2. Look at the list of technical documents in Figure 1-2. How many are you familiar with? Can you think of examples of some of these documents? When would they be likely to be important to you as a reader? Can you think of other types of documents not included in Figure 1-2? Ask some engineering friends how many kinds of documents they have worked on, either as individuals or as part of a group.

3. Think of your own engineering major or specialty. List some engineering fields most closely related to yours, some that are marginally related, and some that are only remotely related. What kinds of technical knowledge do you share with people in these fields? At what point is your common knowledge likely to be no longer useful? What problems can you foresee in communicating technical information with engineers in other fields? What problems would you face if you had to talk about your field to a nonengineering audience?

4. As we point out in this chapter, noise is anything that interferes with efficient transmission of information. We've all experienced noise when trying to communicate with another person—and most of us have at times created it. What kinds of noise do you think you create in your written communication? Is it primarily in your spelling, grammar, sentence structure, organization of thoughts, or what? How about in your spoken communication? What kinds of noise sometimes interfere with your receiving and understanding the written or spoken communication of others?

BIBLIOGRAPHY

Angell, David, and Brent Heslop. *The Elements of E-Mail Style*. Reading, MA: Addison-Wesley, 1994.

Cuevas, Vera. "What Companies Want: The 'Whole Engineer.'" *EE Times Online*. http://www.eetimes.com/salarysurvey/1998/work_companies.html. Accessed February 20, 2003.

Nagle, Joan G. *Handbook for Preparing Engineering Documents*. New York: IEEE Press, 1996.

Naisbitt, John. *Megatrends*. New York: Warner Books, 1982.

Pearsall, Thomas E. *The Elements of Technical Writing*, 2nd ed. Needham Heights, MA: Allyn & Bacon, 2001.

Smith, Edward L., and Stephen A. Bernhardt. *Writing at Work: Professional Writing Skills for People on the Job*. Lincolnwood, IL: NTC Publishing Group, 1997.

2

Some Guidelines for Good Engineering Writing

Complex technical writing is likely to be very difficult to read. Readability further decreases when the writer does not define major ideas for the reader and when the written document is not relevant to the reader's experience and interests. These two impediments can be eliminated if you clearly define your purpose and your audience.

> Ruth Savakinas, "Ready, Aim—Write!" *IEEE Transactions in Professional Communication* 31, no. 1 (March 1988), p. 5

The next two chapters present guidelines for writing efficiently and producing useful documents. Although different people approach writing tasks in somewhat different ways, these guidelines in general follow the overall process used by successful engineering writers and include all the factors you should consider from the time you first face a writing task to the point at which you have a final draft you can be proud of. We have also focused on these topics because they represent common problems you as an engineer are likely to face in the course of writing and formatting your documents.

FOCUS ON WHY YOU ARE WRITING

Before starting to write, you should have a good idea of precisely what you want to communicate to your audience. If these goals aren't first defined in your own mind, you can't really expect your readers to get a clear message. Having this sense of purpose as you write may not guarantee your readers will receive a noise-free message, but writing without a clear goal will almost certainly result in poor communication. Thus, whether you have to write a short memo or a lengthy technical report, you should start with a firm sense of purpose so you can (1) present appropriate supporting data, (2) test its adequacy, and (3) discard anything that is not needed.

Broadly speaking, the purpose of most technical writing is either to present information or to persuade people to act or think in a certain way. Frequently, however, your documents will have to be both informative and persuasive. To fine tune your sense of purpose before writing, ask yourself the following questions.

Do I want to

1. *Inform:* Provide information without necessarily expecting any action on the part of my reader(s)?

2. *Request:* Obtain permission, information, approval, help, or funding?

3. *Instruct:* Give information in the form of directions, instructions, procedures, or the like, so my readers will be able to do something?

4. *Propose:* Suggest a plan of action or respond to a request for a proposal?

5. *Recommend:* Suggest an action or series of actions based on alternative possibilities that I've evaluated?

6. *Persuade:* Convince or "sell" my readers, or change their behavior or attitudes based on what I feel to be valid opinion or evidence?

7. *Record:* Document for the record how something was researched, carried out, tested, altered, or repaired?

How you write any document should be guided by what you want your audience to do with your information, and what they need from the document to be able to do it. Thus, your audience plays a defining role in determining how you approach your task. Do they need to be informed, instructed, dissuaded, warned, encouraged, or what? Only a careful analysis of your purpose or purposes for writing and the nature of your audience can give you the answers and thus enable you to write to the point.

FOCUS ON YOUR READERS

If you found yourself in a remote region and met people who had never seen anything electronic, you wouldn't hand them your scientific calculator or MP3 player and expect them to use it. First, a great deal of technology transfer would have to take

Figure 2-1 You will deal with many different people as your career progresses, so it is best to have a clear picture of who your audience is before beginning to communicate with them.

place; you would have to teach your "audience" how to use your gadget (assuming they cared to know). This may seem obvious, but a lot of technical writing fails because writers make inaccurate assumptions regarding the people who read their documents. Engineers often write without taking adequate time initially to consider the nature, needs, interests, levels of expertise, or possible reactions of those who must read their work. Since you will be writing for many different audiences during your career, as Figure 2-1 illustrates, it is well worth taking the time to think about your audience before writing to them.

Audience analysis is not just a question of being polite, thoughtful, or sensitive. Because your goal is to send a clear, noise-free message through your document to your audience, you must consider their abilities and expectations as you plan, write, and revise.

As an engineer, you may find yourself writing to a variety of people either in your immediate group, close by in the company, elsewhere in the company, or outside the company. Sometimes you will write to your professional and technical peers, sometimes to your superiors, and other times to those "below" you. In all these writing situations, inadequate audience analysis will inevitably result in noise, since different readers need different kinds of data from you.

No matter who you write to, you write because you expect some kind of resulting action—even if it is only nonphysical "action" such as permission, understanding, or a change of opinion. To get results, your communication must bridge a gap between you and your target audience. In the working world this gap is likely to be caused by variations in *knowledge*, *ability*, or *interest*. Obviously, the three may overlap, but to determine where you stand before putting any effort into writing, first identify who your audience is and then ask yourself the following questions.

Knowledge

- Are my readers engineers in my field of expertise who are seeking technical information, and will they be offended or bored by elementary details?

- Are they engineers from a different field who will need some general technical background first?

- Are they managers or supervisors who may be less knowledgeable in my field but who need to make executive decisions based on what I write?

- Are they technicians or others without my expertise and training but with a strong practical knowledge of the field?

- Are they nonexperts from marketing, sales, finance, or other fields who lack engineering or technical backgrounds but who are interested in the subject for nonengineering reasons?

- Are they a mixed audience, such as a panel or committee made up of experts and laypeople?

Ability

- Am I communicating technical information on a level my audience can use?

- Am I using appropriate vocabulary, examples, definitions, and depth of detail?

- Am I expecting more expertise, skill, or action from my audience than I can reasonably expect?

Interest

- Why will my audience want to spend time reading this document?

- Does my document provide the right level of detail and technology to keep my audience's interest without losing them or boring them?

- What is their current attitude likely to be—positive, neutral, or negative?

- Will my document give them the information they want?

The answers to these questions will increase your awareness of the multiple decisions and choices to be made as you plan, write, or revise your document. Remember, in order to deliver a clear message, you should first assess your audience. You need to know who you are writing to and have a clear idea of their technical knowledge, expectations, and attitude toward the subject. If you properly analyze these issues and address them in your document, you are well on your way to communicating effectively.

SATISFY DOCUMENT SPECIFICATIONS

Before writing, you should be aware of any specifications your document must meet. Many audiences expect documents they receive to be within certain parameters. If management asks for a brief memo, they may be irritated when you overload their circuits with a lengthy, detailed treatise. When a technician requests the specs on a frequency tester, he or she won't appreciate a flowery prose discussion on the strengths and weaknesses of the equipment. If you respond to an RFP (request for proposal) that calls for a proposal of no more than 10 pages but submit something twice that long, chances are your proposal will be eliminated from the competition.

Various document specifications exist. Such specifications may require you to provide sections addressing certain topics in your report, such as experimental problems, environmental impact, decisions reached, budget estimates, and so on. The editors of an engineering journal may put limits on the number of words and the number of graphics your technical paper can include. A word limit is frequently placed on the length of an abstract or summary as well as on other sections of a document. Here is the final requirement for a proposal to obtain a research grant from one large funding organization:

> Also required is a nontechnical summary (250 words or less) of the research proposed, expressing significance attached to the project and reasons for undertaking it. This summary will be used for public information and should be written in terms which nonscientists can easily understand.

Many reports have specifications that include requirements not only for their length but also for such matters as headings, spacing, and margin width. Some government agencies, for example, require that the proposals they receive be written in specific formats, in certain fonts, and even with restrictions on how many letters are permitted in each line of text. Here is an example from an RFP for a government research program:

Each proposal shall consist of not more than five single spaced pages plus a cover page, a budget page, a summary page of no more than 300 words, and a page detailing current research funding. All text shall be printed in single-column format on 8-1/2 × 11-inch paper with margins of at least 1 inch on all sides.

Knowing precisely what is expected of you *before* you begin to write will prevent wasted time and give your document a better chance at success.

GET TO THE POINT

Anyone reading your memos, letters, and reports is likely to be in a hurry. Few engineers have the leisure for "biblical" reading—where one reads from Genesis to Revelation to discover how things turn out. Just as your sentences need to be direct, your documents need to have the most important information at the beginning. This means moving from the general to the specific. Readers would much rather know your key points, complaints, requests, conclusions, or recommendations before they read supporting details. For instance, if you did a series of tests to determine whether some equipment should be replaced, your supervisor will want to know what you have found out and what you recommend. A complete, detailed description of your test procedures may be necessary to support your main points and will likely be verified by others—but it could go unread by those in management who need only the bottom line.

Where you tell your readers what they most need to know depends on the kind of document. In a letter, it will be in the opening sentences. In a memo, you should provide a subject line making more than just a vague reference to the overall topic. Look at these examples:

Vague: *SUBJECT: Employee safety*

Better: *SUBJECT: Need for employees to wear hard hats and safety glasses*

Vague: *SUBJECT: Emergency requisitions*

Better: *SUBJECT: Recommendations to change the procedures for making emergency requisitions*

Most memos are now sent by email, which may limit the number of characters for your subject heading. In this case the challenge is to get as much meaning as possible into a small space and to clearly state your key message in the opening sentences of the memo.

In a longer report, your main points should become quickly evident to your reader through an informative title followed by a summary of your findings, conclusions, recommendations, results, or whatever the important information is. (See the chapters on individual reports and the sections on abstracts and executive summaries in this book.) No matter what kind of document you are producing, however, first determine your audience and purpose, and then give your readers the information they most need in the place they can most efficiently access it—the beginning of the paper, rather than buried somewhere in the middle or at the end.

PROVIDE ACCURATE INFORMATION

Even the clearest writing is useless when the information it conveys is wrong. If you state that an ampere is defined as a coulomb of charges passing a given point in 10 seconds rather than 1 second, you have presented wrong information. If you refer to data in Appendix B of your report when you mean Appendix D, the error could stump your readers and cause them to lose confidence in your report.

Inaccurate references to the work of others also will cause your readers to be highly suspicious of the reliability of your entire report—and even of your honesty as a writer (see Chapter 11). Inaccurate directions in a set of instructions or procedures might be frustrating at best, disastrous at worst. Considerable problems have resulted when engineers gave measurements in standard units that were assumed to be metric by others. Another kind of inaccuracy might be a claim that is true sometimes but not under all conditions, for example, that water always boils at 100°C. What about purity and variations in atmospheric pressure?

There is also a great difference between fact and opinion. A *fact* is a dependable statement about external reality that can be verified by others. An *opinion* expresses a feeling or impression that may not be readily verifiable by others. The danger comes when opinions are stated as facts. Note the difference between these two statements:

Fact:	*The NR-48 tool features multiple programmable transmitters and a five-station receiver array.*
Opinion:	*The NR-48 is by far the best piece of equipment for our purposes.*

The second statement might be "correct" but is still only an opinion until supported by verifiable facts. To be strictly honest, the writer should identify it as an opinion unless evidence is presented to support it as fact. In short, make sure that (1) your facts are correct when you write them down and (2) your opinions are presented as such until adequate evidence is provided to verify them.

PRESENT YOUR MATERIAL LOGICALLY

Not only should it be easy to access your document's essential message, but also all your information should be in the right place. This means you must organize your material so that each idea, point, and section is clearly and logically laid out within an appropriate overall pattern. If you are following document specifications provided by someone else, you have little choice but to follow those specs, but even within a prescribed plan of organization you may have some leeway to present material the way you feel is most effective.

As always, think before writing, and keep your readers firmly in mind. If they want to know what progress you have made on a project, what you did on a trip, or how to carry out a procedure, obviously they will expect your material to be in *chronological* order. If they are expecting a description of a piece of equipment or of the layout of some facilities, they should be provided with a description that logically moves from *one physical point to another.*

On the other hand, if you have a number of points to make, such as five ways to reduce costs or six reasons why a project must be canceled, present those points from the *most to the least important*, or vice versa. Perhaps your material needs to be presented in order of *familiarity or difficulty*, as when you are writing a tutorial or textbook. Or you may want to move from the *general to the specific*, as when you write a memo first stating that more stringent safety regulations are needed at your plant and then providing examples of current unsafe practices. Note that Table 9-1 on how to organize material really applies to written material just as much as it does to oral presentations.

MAKE YOUR IDEAS ACCESSIBLE

Without even reading a word, we can look at the pages of a document and get a good idea of how efficiently the material is presented. This impression comes from the structure of the material—specifically, how well the material is laid out in accessible "chunks" for the reader. The two most important factors here are (1) the subdivision of material into sections and subsections with hierarchical headings and (2) paragraph length.

HIERARCHICAL HEADINGS

Even in short engineering documents, a system of headings is essential to keep your material clearly organized and to let readers know what is in each section of the document. Headings and subheadings are also signposts that help a reader get through a report without getting lost. Moreover, they reveal the hierarchical relationships of your material, enabling readers to understand the various levels of detail or importance in your work. Clear and informative headings also give your document good "browseability"; that is, they help readers quickly find the parts of your report that interest them most.

Although practice differs among engineers and organizations, a common format for the first three levels of headings is as follows:

FIRST-LEVEL HEADING

Write first-level headings in capital letters and put them flush with the left margin of the page. Use boldface to make the heading stand out, and separate it from the written material above and below it by at least one space, as in this illustration.

Second-Level Heading

Also place second-level headings flush with the left margin with at least one space separating them from any text. Capitalize only the first letter of each main word, and make these headings boldface. If you don't like boldface type, you can underline your headings, although underlining clutters the text. In any case, don't use both boldface and underlining for headings.

Third-level headings. Place third-level headings on the same line as the text they precede. They are capitalized as a sentence would be and can be in boldface or italics.

Note Each level of heading after the first can be indented two or three spaces for visual effect if you wish. The accompanying text would then also be indented with the heading. For an example, see Figure 11-1 of this book.

NUMBERED HEADINGS

Sometimes you may be required to add a numbered, or decimal, system to your headings; many companies and suppliers require such numbering. A number system gives readers easier reference to parts of a very long report. Note that these different levels of headings can also be successively indented, although many companies don't follow this practice.

FIRST LEVEL   **1. 0 QUALITY ASSURANCE PROVISIONS**

Second Level   **1.1 Contractor's Responsibility**

Third Level   **1.1.1 Component and material inspection**

Fourth Level   **1.1.1.1 Laminated material certification**

When you use this system, make sure it doesn't get out of control. If your material is so complicated or detailed that you are getting down to levels such as 2.11.3.4.6.23, as some manuals do, then maybe it's time to inspect your document closely to see where you can break it up into smaller, more manageable sections or short chapters, each with its own verbal heading and independent hierarchies within it.

These structural elements of a document (and again, it doesn't matter whether it's a 2-page memo or a 500-page manual) can be planned ahead of time. Writing skills aren't needed so much for this as planning and outlining skills—plus an awareness that the headings, divisions, and subdivisions in your document play a vital part in making your information clear and easily available to your readers. So spend some time thinking about how you're going to arrange and format your document; this will help you avoid noise at the structural level before you even begin to write. You might, of course, want to further improve the structure and organization of your paper after completing the first draft. Word processing software now makes this easy and even enjoyable.

PARAGRAPH LENGTH

No one, especially in technical fields, wants to read a solid page of wall-to-wall text of difficult material. A busy manager, for example, will want to absorb your information in as easily digestible pieces as possible.

Dense text on a page creates noise simply because it's so discouraging. When your readers are trying to follow demanding technical information, they are already challenged, and presenting it to them in solid page-long chunks is going to give them mental indigestion. Later, if they want to quickly find a point you made or a piece of data you presented, they are going to have trouble locating it if they have to wade through a ponderous paragraph to get to it.

A rule of thumb in technical writing states that paragraphs should not be much over 12 lines long, but it's better if they are even shorter in general. Occasionally you will have to go over the 12-line rule, but try not to do so too often. When editing your work, look for any overly long paragraphs and try splitting them into two—and when you do, remember that you may have to add a transitional word or phrase.

As an illustration of what we're getting at, we have reformatted most of this section on paragraph length with no paragraph breaks:

No one, especially in technical fields, wants to read a solid page of wall-to-wall text of difficult material. A busy manager, for example, will want to absorb your information in as easily digestible pieces as possible. Dense text on a page creates noise simply because it's so discouraging. When your readers are trying to follow demanding technical information, they are already challenged, and presenting it to them in solid page-long chunks is going to give them mental indigestion. Later, if they want to quickly find a point you made or a piece of data you presented, they are going to have trouble locating it if they have to wade through a ponderous paragraph to get to it. A rule of thumb in technical writing states that paragraphs should not be much over 12 lines long, but it's better if they are even shorter in general. Occasionally you will have to go over the 12-line rule, but try not to do so too often. When editing your work, look for any overly long paragraphs and try splitting them into two—and when you do, remember that you may have to add a transitional word or phrase. Some of your paragraphs will be much shorter than 12 lines, of course, especially if they are transitional paragraphs or convey particularly complex material. If you are writing a manual or set of procedures, most "paragraphs" will probably be one-sentence directives such as *Move the pointer to the next slide and click again.* One last caution on paragraphs: Try to avoid "orphan lines" in your document—paragraphs for which the first sentence begins on the last line of a page, or the last sentence appears at the top of a page.

Looking at this word mass, you can appreciate the need to make your information accessible by presenting it in fairly short "chunks" of information.

Some of your paragraphs will be much shorter than 12 lines, of course, especially if they are transitional paragraphs or convey particularly complex material. If you are writing a manual or set of procedures, most "paragraphs" will probably be one-sentence directives such as *Move the pointer to the next slide and click again.*

One last caution on paragraphs: Try to avoid "orphan lines" in your document—paragraphs for which the first sentence begins on the last line of a page, or the last sentence appears at the top of a page.

USE LISTS FOR SOME INFORMATION

A well-organized list is sometimes the most efficient way to communicate information. If you have to present steps in a procedure, materials to be purchased, items to be considered, reasons for a decision, or groceries to be bought, a list might well be the best way to go because readers retrieve some kinds of information from a list more easily than from a passage of prose. Look at the following:

> First of all, set the dual power supply to +12V and −12V. Next, set the op-amp up as shown in Figure 1. Use a 1 Vpp sinewave at 1 kHz and then plot the output waveform on the HP digital scope. Then obtain a Bode plot for the gain from 200 Hz to 20 kHz.

You could present this information more efficiently in list form:

> 1. Set the dual power supply to +12V and −12V.
> 2. Set the op-amp up as shown in Figure 1.
> 3. Use a 1 Vpp sinewave at 1 kHz.
> 4. Plot the output waveform on the HP digital scope.
> 5. Obtain a Bode plot for the gain from 200 Hz to 20 kHz.

There are three main types of lists: numbered lists (as just shown), checklists, and bulleted lists. You can combine these in various ways to get sublists if you wish. Use a *numbered list* to indicate when a set of data follows a certain order, as in the previous example. Numbered lists can also be used to indicate an order of importance in your data, such as a list of priorities or needed equipment.

Sometimes lists are formed using upper- or lowercase letters in alphabetical order. Numbers are usually best for the main entries in your list, however, since most people are more familiar with moving through steps 1 to 10 than steps (a) through (k). You can always consider using letters for sublists.

Checklists can be used to indicate that all the items on your list must be tended to, usually in the order presented:

> ☐ Connect the monitor to the computer through the monitor port.
> ☐ Connect the keyboard and mouse to the computer through the ASF port.
> ☐ Connect the power supply to the computer.
> ☐ Connect the printer to the printer port.
> ☐ Connect the modem to the modem port.

These instructions could also be presented thus:

1. Connect the monitor to the computer through the monitor port. ☐
2. Connect the keyboard and mouse to the computer through the ASF port. ☐
3. Connect the power supply to the computer. ☐
4. Connect the printer to the printer port. ☐
5. Connect the modem to the modem port. ☐

When checklists get longer than ten boxes, try to break them down into smaller, more manageable sections, and give each section its own subheading.

Bulleted lists are commonly used when items in the list are in no specific order, as in the following example.

Some of the main concerns of environmental engineering are
- Air pollution control
- Public water supply
- Wastewater
- Solid waste disposal
- Industrial hygiene
- Hazardous wastes

Today's word processing software allows you to create bullets easily, and to substitute arrows or tick marks if you wish. Lengthy bulleted lists—over seven items—are hard for readers to refer to, so use numbers for longer lists even if no order of priority is intended.

PUNCTUATION AND PARALLELISM IN LISTS

If the lead-in to your list ends with a verb, no colon is necessary. *The five priorities we established are* would *not* require a colon after *are* because the list is needed to logically and grammatically complete the statement. (Also see the bulleted list example shown earlier.) A lead-in like *We have established the following five priorities* would be followed by a colon, however, because the statement is grammatically complete.

If the items in your list are complete sentences and contain internal punctuation, you should conclude each one with a period. Otherwise, a period at the end of list items is optional. Capitalizing the first listed item is up to you, unless each entry is a complete sentence. Whichever style of punctuation and capitalization you use, be consistent.

Another concern when writing lists is to maintain "grammatical parallelism" between entries. This simply means that if some entries begin with a verb, all entries should do so; if most begin with a noun, all should. This makes for smoother reading and logical neatness. Note how the following list is bumpy due to problems with parallelism:

Last week we accomplished the following for WW3-a:

- Completed BIU, ICACHE, and ABUS logic design.
- All instruction buffer blocks have had final simulations.
- Written and debugged 75% of test patterns.
- Scheduling of first silicon reticules for WW4-a with Vern Whittington in Fab 16.

Making the items in the list parallel cuts out some psychological noise:

Last week we accomplished the following for WW3-a:

- Completed BIU, ICACHE, and ABUS logic design.
- Ran final simulations on all instruction buffer blocks.
- Wrote and debugged 75% of test patterns.
- Scheduled first silicon reticules for WW4-a with Vern Whittington in Fab 16.

FORMAT YOUR PAGES CAREFULLY

In addition to how you divide information up and how long you make your paragraphs, other factors can have a positive or negative effect on your reader. People prefer print that is visually accessible and pleasing. You can create psychological noise if you fail to meet these preferences, but you can easily prevent it by keeping the following pointers in mind.

MARGINS

Leave ample margins around your text to help prevent your pages from appearing overloaded. Standard margins are one inch all around your page, but it is possible to go a little above or below this if you have to. Make sure the margins are consistent on all pages. If you can, let your lines of text wrap around with a "ragged" right-hand margin rather than aligning them on the right, since this makes for easier reading. If your report is important enough to be bound like a book, you will need a wider-than-usual left margin to accommodate the binding and ensure that the first word or so of each line is still readable.

TYPEFACE

Typeface is the style of individual letters and characters. *Serif* and *sans (without) serif* are the two general typefaces, with serif fonts having small strokes or stems on the edges of each letter. Books, magazines, and newspapers generally use serif fonts for their text, so this is what people are most used to seeing. Sans serif fonts can be effective for titles and headings, but serif fonts make larger quantities of text more readable because the little stems bind the letters and guide the reader's eye from letter to letter.

Sans serif: The electric car prototype has regenerative braking, which recharges the supply while decelerating the vehicle.

Serif: The electric car prototype has regenerative braking, which recharges the supply while decelerating the vehicle.

Standard type size is 10 to 12 points. You should use larger or smaller sizes only for special effect in titles, captions, warnings, and such. Generally avoid sentences with all capital letters because in a long sequence of uppercase letters you have the same visual contours, making such a sentence somewhat more difficult to read:

THE GOVERNMENT PLANS TO ESTABLISH A HIGH-LEVEL ADMINIS-TRATIVE COUNCIL TO COORDINATE SCIENCE AND TECHNOLOGY.

Capitalized words should be used to emphasize a heading or directive, however:

DANGER: A 7000 V potential exists across the transformer output terminals.

WHITE SPACE

White space refers to areas of a page not filled with text or graphics. When reading, we tend to take white space for granted, but it plays an important part in a document by creating a path for a reader's eyes, isolating and emphasizing important data, and providing "breathing room" between blocks of information. Thus it can have a posi-tive effect by making difficult technical material appear more accessible and less threatening.

You will have enough white space on your pages if you do the following:

- Provide adequate and consistent margins.
- Leave a space between all paragraphs.
- Leave spaces before and after every heading and subheading.
- Leave one or two spaces between text and graphics or lists.
- Leave a space before and after each equation in the text.
- Indent subheadings or text where appropriate.
- Use a ragged (unjustified) right margin.

Much of what we advise to make sure your documents are well formatted, and thus visually accessible to your readers, is illustrated in Figure 11-1. Shown as an example of effective documentation, the page is also well formatted. Notice how sec-tions are clearly organized and labeled, using both numeric and verbal hierarchical headings and subheadings. The page is not cluttered or dense, and the prudent use of spacing creates a page that is not daunting to a busy reader.

EXPRESS YOURSELF CLEARLY

Engineering is considered a precise discipline—although in reality, as most engi-neers will admit, it's not always as precise as we would like it to be. Machine parts, for example, may be allowed a certain degree of variation or tolerance within a spec-ified zone and still be interchangeable. Similarly, you have some choices in how you express yourself in engineering writing. In English, you can often say the same thing three or four different ways, but your overriding concern should always be to state what you have to say clearly and to the point. Don't force your readers to work harder than necessary to grasp what you have written; your sentences must convey a single meaning with no room for interpretation or misunderstanding. If your readers yearn for uncertainty and suspense, they can read a romantic novel or detective story, and if they enjoy different connotations and levels of meaning, they can read poetry. The following sections discuss some pitfalls to avoid and some goals to strive for.

AMBIGUITY

The word *ambiguous* comes from a Latin word meaning to be undecided. Ambiguity primarily results from permitting words like *they* and *it* to point to more than one pos-sible referent in a sentence, or from using short descriptive phrases that could refer to two or more parts of a sentence. In either case, your reader becomes confused—undecided—and may interpret your sentence differently than you intended, as illustrated in the following example.

Ambiguous: *Before accepting materials from the new subcontractors, we should make sure they meet our requirements.*

(Who are *they*, the materials or subcontractors?)

Clear: *Before we accept them, we should make sure the materials from the new subcontractors meet our requirements.*

Ambiguous: *The microprocessor interfaced directly with the 7055 RAM chip. It runs at 5 MHz.*

(What does *it* refer to?)

Clear: *The microprocessor interfaced directly with the 7055 RAM chip. The 7055 runs at 5 MHz.*

Ambiguous: *Our records now include all development reports for B-44 engines received from JPL.*

(What was received from JPL—the reports or the engines?)

Clear: *Our records now include all B-44 engine development reports received from JPL.*

Ambiguous: *After testing out at the specified high temperatures, the company accepted the new chip.*

(Did the company or the chip test out at the high temperatures?)

Clear: *The company accepted the new chip after it tested out at the specified high temperatures.*

VAGUENESS

If ambiguity causes readers to see more than one meaning in your writing, vagueness causes them to see no useful meaning at all. What would you think if your doctor told you to "take a few of these pills every so often"? You would want him or her to provide some facts and figures. Explanations or directions lacking specific detail sound fuzzy and unfocused, more like personal opinion than useful data.

Abstract words are not inherently wrong, but they fail to provide the precision effective technical writing needs. Try to avoid abstract words and phrases such as *pretty soon, substantial amount,* and *corrective action,* and replace them with terms that have exact meaning, such as *in three days, $8,436.00,* and *replace the altimeter.* Here are two more examples of vague writing and ways they can be remedied:

Vague: *The Robotics group is several weeks behind schedule.*

Useful: *The Robotics group is six weeks behind schedule.*

Vague: *The CF553 runs faster than the RG562 but is much more expensive.*

Useful: *The CF553 runs 84% faster than the RG562 but costs $4320–$2840 more than the CF553.*

As you can see in the second example, vague writing might require fewer words, but it's rarely wise to be concise at the expense of precision. This is especially true when writing instructions and specifications.

On the other hand, vagueness can be an asset to people who don't want to reveal too much—or who have nothing to reveal because they've done nothing. The following satirical "Progress Report for All Occasions" has been going around industry for some years now and is a monument to vague writing:

During the report period that encompasses the organized phase, considerable progress has been made in certain necessary preliminary work directed toward the establishment of initial activities. Important background information has been carefully explored and the functional structure of components of the cognizant organization has been clarified.

The usual difficulty was encountered in the selection and procurement of optimum materials, available data, experimental data, and statistical analysis, but these problems are being attacked vigorously, and we expect that the development phase will continue to proceed at a satisfactory rate.

You might want to write something like this—if in reality you have no progress to report!

COHERENCE

The root of the word *coherence* is *cohere,* meaning to stick together, and as you know, a cohesive does just that. Coherence in writing refers to how well paragraphs and even complete documents "stick together"—that is, stay focused on their true subject. In a coherent paragraph, all the sentences clearly belong where they are because they address only the topic of the paragraph and are logically connected to one another. You might say the sentences stick to the point and stay there. Coherence in a complete report also means how well the report is designed to take the reader through its paragraphs and sections by means of clear transitions such as headings and subheadings, and how all the sections focus on and support the subject of the report. (Our chapters on report writing will show you how to achieve coherence in longer documents.)

You can achieve coherence in your paragraphs by making sure each sentence clearly relates to the one before it and after it. This means opening with your main point or topic sentence, repeating key words where needed, and using transitional words (see Chapter 3) and pronouns to link sentences as they build up the paragraph. Note how the following paragraph lacks coherence and how it is improved by the devices in boldface in the revised version.

Poor Coherence

A significant disadvantage of the 125-H CRT is its high power consumption. The tube requires substantial power to produce the high voltages and currents that are necessary to drive and deflect the electron beam. The 125-H is inefficient—only about 10% to 20% of the power used by the tube is converted into visible light at the surface of the screen. The 125-H is poorly suited for portable display devices that run on batteries, where lower power consumption is necessary. We should consider other options before committing to purchase the 125-H.

Effective Coherence

A significant disadvantage of the 125-H CRT is its high power consumption. **This** *tube requires substantial power to produce the high voltages and currents that are necessary to drive and deflect the electron beam.* **In addition,** *the 125-H is inefficient—only about 10% to 20% of the power used by the tube is converted into visible light at the surface of the screen.* **Thus,** *the 125-H is poorly suited for portable display devices that run on batteries, where lower power consumption is necessary.* **Because of this drawback**, *we should consider other options before committing to purchase the 125-H.*

Directness

Being as direct as possible in your writing lets your reader grasp your point quickly. Suspense might be thrilling, but a busy technical reader wants access to your information quickly and easily. The most important part of your message should come at the beginnings of your sentences and paragraphs. Here are some examples of what this means at the sentence level:

Indirect: *After a long and difficult development cycle due to factory renovation, the infrared controller will be ready for production in the very near future.*

Direct: *The infrared controller will be ready for production March 4. Its development cycle was slowed by the factory renovation.*

Indirect: *Fred has been busily working on this project. This past week he also reworked the logic diagrams, rewired the controller arm, and redesigned all of the RIST circuitry.*

Direct: *Fred redesigned the RIST circuitry on Thursday. He also reworked the logic diagrams and rewired the controller arm last week.*

USE EFFICIENT WORDING

Opinions vary on how much it costs a company for an employee to produce one written page of technical information, but as stated in Chapter 1, it can be anywhere up to and beyond $200. When you think of all the people writing letters, memos, reports, manuals, proposals, and countless other documents for industry, you see how the costs mount up. Add to this the fact that most of us have little training in producing concise prose, and you can appreciate how sharpening your writing and editing skills can mean not only saving time, but money. Moreover, since we all tend to be wordy, carefully editing our work can often reduce or eliminate a lot of time-consuming work for our readers.

WORDINESS

Using an unnecessarily pompous word instead of a straightforward one can cause your readers to slow down. Choose the simplest and plainest word whenever you can. Your readers can be distracted or even confused by words that call attention to themselves without contributing to meaning. This pitfall becomes even more likely if some of your readers are not native speakers of English, as is often the case in engineering fields today. Write to communicate rather than to impress, or as the saying goes, "Never utilize *utilize* when you can use *use*."

A few of the more ostentatious—oops, make that *showy*—words found in engineering writing are listed here, with some plain, equally efficient counterparts:

commence	*start*	fabricate	*make*	proceed	*go*
compel	*force*	finalize	*end*	procure	*get*
comprises	*is*	initiate	*begin*	rendezvous	*meet*
employ	*use*	optimal	*best*	terminate	*end*
endeavor	*try*	prioritize	*rank*	visitation	*visit*

Wordiness can also result from using far more words than you need to express an idea. Unkind editors sometimes refer to this as *verbiage* (by analogy to garbage?). Few of us appreciate hearing

> *I regret to say that at this point in time I basically do not have access to that specific information.*

when a simple "I don't know" is enough. Similarly, your reader is unlikely to thank you for having to plow through

> *It is our considered recommendation that a new computer should be purchased.*

when you could have simply said you recommend buying a new computer.

You can eliminate a lot of wordiness in your writing by training yourself to edit carefully and to make every word count. Look at the following three pairs; you will see which sentences are more efficient and noise free.

> *It is essential that the lens be cleaned at frequent intervals on a regular basis as is delineated in Ops Procedure 132-c.*
>
> Clean the lens frequently and regularly (see Ops Procedure 132-c).
>
> *The location of the experimental robotics laboratory is in room 212A.*
>
> The experimental robotics lab is in 212A.
>
> *There are several EC countries that are now trying to upgrade the communication skills of their engineers.*
>
> Several EC countries are trying to upgrade the communication skills of their engineers.

You can also reduce wordiness by avoiding certain pretentious phrases that have unfortunately become common. A good stylebook will give numerous examples, but here are a few that crop up frequently in engineering writing:

Verbiage	Efficient
a large number of	*many*
at this point in time	*now*
come in contact with	*contact*
exhibits the ability to	*can*
in the event of	*if*
in some cases	*sometimes*
in the field of	*in*
in the majority of instances	*usually*
in the neighborhood of	*about*
in view of the fact that	*because*
in view of the foregoing	*therefore*
serves the function of being	*is*
subsequent to	*after*
the reason why is that	*because*
within the realm of possibility	*possible*

Check your writing for such unnecessary phrases and for unneeded words in general—as we do in the next sentence. You may ~~often~~ find t~~hat there are a number of~~ words ~~contained in your writing~~ that can be ~~safely~~ eliminated without any ~~kind of~~ danger to your meaning ~~whatsoever~~.

Note If you let your writing "cool off" for a while and come back to edit later, chances are you will discover more wordiness than if you try to edit immediately after writing.

REDUNDANCY

One category of verbiage is redundancy. This means using words that say the same thing, like *basic fundamentals*, or phrases that duplicate what has already been said, as in *They decided to reconstruct a hypothetical test situation that does not exist.* In fact, if you master the art of redundancy, you can make everything you write almost twice as long as need be. A few common redundant pairs are identified in the following list, but the list is far from exhaustive.

Again, we all can be wordy at times, so it's a good idea to edit your writing once simply looking for redundancy and wasted words. Grammar-checking software can help, but you still need human editing to remove this kind of noise from your writing.

Redundant	Efficient
alternative choices	*alternatives*
actual experience	*experience*
completely eliminate	*eliminate*
component part	*component (or part)*
connected together	*connected*
collaborate together	*collaborate*
diametrically oppose	*oppose*
exactly identical	*identical*
integral part	*part*
just exactly	*exactly*
permeate throughout	*permeate*
prove conclusively	*prove*
rectangular in shape	*rectangular*
12 noon	*noon*
very best	*best*

TURNING VERBS INTO NOUNS

Replacing a perfectly good verb (action) with a noun (the name of an action) is unfortunately common in much engineering writing. This is often the result of wanting to write in the passive rather than active voice. Look at these pairs of sentences:

An analysis of the data will be made when all the results are in.

We will analyze the data when all the results are in.

An investigation of all possible sources of noise was undertaken.

All possible noise sources were investigated.

Acknowledgement of all incoming messages is performed by the protocol handler.

The protocol handler acknowledges all incoming messages.

It's easy to see which sentences are shorter and more natural.

If you take the verb that really matters in a sentence (such as *analyze*, *investigate*, and *acknowledge* in these examples) and make a noun of it, you are forced to add another, generally weaker, verb to convey your meaning. Thus, you will write *made a selection of* instead of *selected*, or *procurement of services can be accomplished by* instead of *services can be procured by*. Note that many such verbs when changed into nouns need to be followed by *of*. Grammar checkers use this as a cue to warn you of the problem, but again, there is no better tool than your own editing skills—or those of a competent and honest colleague—to free your writing of verbiage.

MANAGE YOUR TIME EFFICIENTLY

Few engineers feel they have enough time to do the writing required of them. Often a memo is hastily churned out or a report is rapidly thrown together and tacked on the tail end of a project. As with anything done in a hurry, the results are usually not the best. As the pressure to get a piece of writing out increases, error—that is, noise— also increases. Rather than leaving your writing to the last minute, it is far better to consider it just as much a part of your professional activities as designing, building, and testing.

FINDING AND USING TIME

There are a number of ways to find time to spend on careful writing and editing, but most are not too attractive. You can get to work an hour earlier, or take work home at night (plenty of successful engineers do). You can use your breaks to get away from distractions and concentrate on your writing tasks. You might designate a specific time each day as your writing period—if your colleagues and other duties permit this. You can write on your laptop computer at airports, in flight, on trains, in hotels, or in waiting rooms.

However, as stated earlier, it's much more practical to make your written work an organic part of your daily schedule. In this way you can assign brief time periods to write short memos and letters or small sections of a report. Larger chunks of time can be designated for concentrating on longer writing tasks.

OUTLINES, DEADLINES, AND TIMELINES

When you have to write anything over two pages long, it's useful to first spend some time making a rough outline. This outline does not have to be set in concrete—that is, you don't have to slavishly follow it once you've written it, and it can be altered at any time—but it will give you some indication of what is involved in producing the finished paper. It will also help you divide your task into smaller sections that can then be written separately at different times, and not necessarily in order. Less

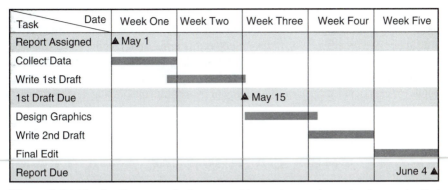

Task Date	Week One	Week Two	Week Three	Week Four	Week Five
Report Assigned	▲ May 1				
Collect Data					
Write 1st Draft					
1st Draft Due			▲ May 15		
Design Graphics					
Write 2nd Draft					
Final Edit					
Report Due					June 4 ▲

Figure 2-2   The timeline you make for your writing project can be as simple or as detailed as you wish. Make sure you have all your important tasks and due dates down, however, and then do everything you can to keep to them.

demanding sections, for example, can be relegated to short periods of available time or to times when more distractions surround you.

Even if a deadline for completing a document hasn't been imposed on you, it's a good idea to establish one for yourself. Estimate how long you expect the job to take, and schedule back from there. You might even draft a timeline for yourself, showing each date by which you should have completed specific parts of the paper. (See Figure 2-2.) Always allow yourself enough time at the end to review and edit the entire document.

EDIT AT DIFFERENT LEVELS

Rather than glance over their finished document once or twice in hopes of randomly finding anything in need of improvement, many writers like to take a more methodical approach to editing. You might want to try this. First, check your document for technical accuracy. Then decide what "writing levels" to approach your editing on, and go through your document at least once on each level.

Level 1 is the nitty-gritty level of mechanics, spelling, punctuation, typos—all the basics we were supposed to master in elementary and high school. Again, a good word processing program will provide you with suggestions on spelling and grammar; however, *you* must make the final choices on many of these options. You might also call upon the services of a friend or colleague who is well grounded in these basics.

Level 2 involves looking at such things as paragraph and sentence length and structure, possible verbiage, and precise word choice. Is the tone of your document appropriate? Have you used the active voice where possible? How about transitions, parallelism, and emphasis where called for?

The final level, level 3, is the more global level of the document, where you check the overall format, organization, and appearance. Is the work arranged the way it should be? Are specifications (if any) followed? Is it the right length? Have you used the best font size, margins, and spacing? Are headings, subheadings, lists, and graphics used effectively and consistently? Is the title page attractive? How about the "packaging" of the document, such as the quality of paper used, the binding, and the covers?

SHARE THE LOAD: WRITE AS A TEAM

Not many engineers write lengthy reports by themselves. Technical people work together as teams for research, design, development, and testing, and often find they must team up to write proposals, manuals, completion reports, and a lot of other technical documents. Team writing is not always easy, especially when people with different degrees of writing ability or ego investment are involved, or when team members are torn between team responsibilities and other duties. If your group plans the team project carefully, however, it can turn out to be relatively painless and very rewarding because as a team you will be tapping into far more knowledge, skill, and creativity than you can bring to a project alone.

A team is a group of two or more people who interact and coordinate their work to accomplish a specific goal. When you work on a team project or help put together a long written document with others, you should be prepared to do the following:

- Communicate
- Coordinate
- Collaborate
- Cooperate
- Compromise

This list might seem obvious, but many teams fail to reach their potential because some members have difficulty in following it. Some people even see *collaboration* and *compromise* in their more negative connotations rather than as the positive attitudes they are meant to be in the context of team activities. Let's look at each one briefly.

- *Communicate.* Obviously, very little teamwork is possible without frank and open communication. This means that members of the team create an atmosphere that enables free discussion at all times. It also means that the channels of communication (i.e., email, telephone numbers, mail addresses, and meeting times and places) are all common knowledge to each member.
- *Coordinate.* Since team members are often scattered when not physically working together, it's very important that everyone knows what the others are doing, who is responsible for what, when the next deadline or meeting is, and

so forth. Often one member of a team is appointed as the coordinator. If that person does the job well, there will be a minimum of frustration, repetition, redundancy, or uncertainty among the team.

- *Collaborate*. The Latin root of this word means "to willingly labor." In a team setting it means just that—to willingly assist one another. In the spirit of collaboration you will, for instance, assist a partner on some work if necessary, or work at understanding what another team member is doing. You will also freely share your own work with the other team members, and work at creating a final document that is unified and seamless.

- *Cooperate*. An attitude of cooperation is essential to the smooth working of any team project. If the project has a designated leader, you will do all you can to cooperate with that person and to accept his or her decisions, deadlines, changes, or reassignments. Such "executive" actions on the part of the leader hopefully will be the result of open discussion with all team members, but there may be times when this is not possible, and you may have to cheerfully accept a decision you have no control over.

- *Compromise*. This word has two meanings, and only one of them is somewhat derogatory. The other meaning refers to making mutual concessions in order to reach a goal. In practical teamwork, this means you may sometimes have to give a bit on an attitude, opinion, approach, method, or course of action, because by doing so you help the team reach its overall objective. Compromise should as much as possible be the outcome of open and friendly communication.

Besides maintaining the attitudes just described, there are three practical ways you can employ teamwork to produce a written document. Some methods work for some groups, others work for other groups. We rarely work in ideal circumstances, and you may have to be flexible when working with others on a writing task. The three methods, from the least preferable (but the most commonly used) to the most effective, are as follows:

1. Divide the length of the assignment by the number of people involved, and get each to write his or her share. Individuals will do any research needed for their own sections and should write and edit them. Then the document can be "glued together."

Unfortunately, this method may not result in a very efficient or effective product. Individuals bring their personal writing style, vocabulary, quirks, and weaknesses to their part, and their material may overlap with other parts of the report or fail to provide important transitions between sections. You will still need a strong writer as "overseer" and final editor who can take the completed draft and mold it into a coherent and useful document.

2. Have one person organize the material, write the entire draft, edit it, and pass the finished product on to the next member of the team. This person will add, delete, rearrange, and re-edit as he or she sees fit. The third

member of the team will do the same, and so on down the line. The assumption here is that when all team members have had their say, the document will be as complete and close to perfect as can be.

With a closely knit and cooperative team this method *might* result in an effective report rather than a total mess, but you will still need a strong document manager/editor to monitor each step in the process. You might even find this system bringing friendships to an abrupt end. Moreover, if team members want to see what others have done to *their* version of the draft, and are inclined to debate and dispute each amendment, you could be a long way past deadline before everything is set right and everyone is satisfied.

3. By far the best way to produce a team document is to assign each member to different tasks according to that member's strengths and interests:

   a. Designate one person as project manager to organize and assign tasks, check that the project is on schedule, and even referee disputes if necessary.

   b. Have another team member get together the needed information for the document, write notes, and put together a very rudimentary draft.

   c. Get the next member, the designated "strong" writer, to generate a working draft of the paper. Ideally, this person is good at writing, enjoys writing—and has read this book.

   d. If possible, get yet another team member with editing skills to act as quality control officer, reading, checking, editing, and in general perfecting the document while working closely with the previous writer.

Using this method, everyone on the team can bring particular strengths to the task and play a significant part in producing the document. Each person has direct access to the document manager, knows what the others' responsibilities are, and has the satisfaction of being uniquely involved in the job. This is the ideal situation. Even with this method you may have to compromise sometimes, double up on tasks, or mix this method with elements of the first two described. Whatever the situation, though, carefully planning and assigning collaborative writing tasks to team members *before* the writing project begins will result in a more efficiently produced document that is both coherent and useful.

EXERCISES

1. Think of some significant communication events you have experienced in the past several months at work or in class. What kinds of audiences were involved? Did a lack of clearly defined audience and purpose cause noise in the communication process? How would a

more complete analysis of the audiences have enabled technical information to be transferred more efficiently?

2. Look inside the back cover of an IEEE or other technical journal, where you will find a page of advice for authors who wish to publish in that journal. To what extent does the information provide specifications for the articles to be published? Are specifications given for such details as abstracts, length, headings, margins, columns, graphics, size of print, references, and so on? If you still have questions about how a paper for that journal should be written and formatted, how would you get in touch with the editor?

3. Go to the library or online and find a government or industry report on a subject that interests you. Who is the assumed audience? Does the report get to the point right away or does it keep you guessing until the end? How useful are the headings and subheadings? Is it easy to outline the plan of organization the author has used? How do divisions and paragraph length add to the accessibility of the information? Could any of the information be better presented in list form? Select three or four random paragraphs and closely analyze them for ambiguity, wordiness, unnecessary technical jargon, and nouns that could be turned into verbs. Then rewrite those passages.

4. Keep a log of the time you spend writing a document. How long did it take you? Were you working under a deadline? How much time each day was spent planning, writing, and editing? Did others have a part in writing the document, and if so, how were tasks or sections delegated? Were you satisfied with the completed document? Was whoever assigned you the task satisfied with your work? What factors would have enabled you to do an even better job?

5. Take some examples of your own recent writing and analyze them in light of each of the guidelines in this chapter.

BIBLIOGRAPHY

Alred, Gerald J., Charles T. Brusaw, and Walter E. Oliu. *The Technical Writer's Companion.* New York: Bedford/St. Martin's, 2002.

Nadziejka, David E. "The Levels of Editing Are Upside Down." *Proceedings of the International Professional Communication Conference*, September 28–October 1, pp. 89–93, 1984.

Nagle, Joan G. *Handbook for Preparing Engineering Documents: From Concept to Completion.* New York: IEEE Press, 1996.

Paradis, James G., and Muriel L. Zimmerman. *The MIT Guide to Science and Engineering Communication*, 2nd ed. Cambridge, MA: MIT Press, 2002.

Pfeiffer, William S. *Pocket Guide to Technical Writing*, 2nd ed. Upper Saddle River, NJ: Prentice Hall, 2001.

Stevenson, Susan, and Steve Whitmore. *Strategies for Engineering Communication.* New York: John Wiley and Sons, 2002.

3

ELIMINATING INTERMITTENT NOISE IN WRITING

Vigorous writing is concise. A sentence should contain no unnecessary words, a paragraph no unnecessary sentences, for the same reason that a drawing should have no unnecessary lines and a machine no unnecessary parts. This requires not that the writer make all his sentences short, or that he avoid all detail and treat his subjects only in outline, but that every word tell.

Strunk and White, *The Elements of Style*, p. 17

There arises from a bad and inapt formation of words, a wonderful obstruction of the mind.

Sir Francis Bacon

Errors that crop up here and there in writing, causing what Bacon called "a wonderful obstruction of the mind," are often referred to as faulty mechanics by English teachers and can be thought of as sporadic or intermittent noise. Such noise occurs here and there on a page, rather than affecting the whole document the way a poor choice of font size or confused organization of material might. Of course, enough intermittent noise in a document, such as repeated misspellings or numerous sentence fragments, can easily turn into constant noise. Such noise will give your reader an impression of hastily and carelessly produced work undeserving of the response or feedback you want.

To help you eliminate intermittent noise, this chapter looks at where it is most likely to occur in spelling, punctuation, sentence structure, and technical usage. We also give some pointers on how to edit your writing to remove occasional noise.

SPELLING AND SPELL CHECKERS

You might think electronic spell checkers have eliminated any need to be a careful speller. Unfortunately, this is not the case. With apologies to Shakespeare, we took his words "A rose by any other name would smell as sweet" and ran them through a spell checker as *A nose by any outer dame wood small as sweat*. No red flags were raised by the program. Nor will spell checkers catch common errors such as confusing *there* for *their*, or *to* for *too*. Some typographical errors simple give you other words that will pass unnoticed, as in this sentence. A very slight slip of the finger on the keyboard can make the difference between asking for some forms to be *mailed* to you or *nailed* to you. A quick transposition could render a memo *nuclear* rather than simply *unclear*.

At best, the effect of poor spelling on your readers is a sense of annoyance, or at least of having their attention distracted by something other than what you want to communicate. At worst, noise created by spelling glitches can bring readers to a stop and cause them to seriously question your ability as a writer. They might even suspect that a person who is careless with spelling could also be inept in more critical technical matters.

To reduce or eliminate any noise in your writing caused by incorrect spelling, use a spell checker but also have a standard dictionary nearby. A current dictionary is the only resource that can reliably answer questions such as the following:

- Whether there is more than one way to spell a word, or what the accepted plural forms of words such as *appendix* or *matrix* are
- How words like *well-known* or *so-called* are hyphenated, or whether a computer is *on-line* or *online*
- Whether it is appropriate to write about *FORTRAN, Fortran,* or *fortran*
- What the difference between British and American spelling or usage might be
- What the accepted past tense is of recent verbs that have come into technical English, such as *input*

It is especially important for an engineer to use a current dictionary. English is a dynamic language, and the language of science and technology changes even more rapidly as knowledge increases and devices are developed. You won't find words like *software*, *modem*, and *LED* in a dictionary from the 1950s, and since then older words such as *bug*, *hardware*, *interface*, and *mouse* have taken on new meanings. Some usage has yet to be decided on: Would a computer shop advertise that it repairs *mice* or *mouses*? Do you send *e-mail, E-mail*, or *email*? (As of now all three options are still used, but *email* seems to be winning.)

PUNCTUATION

Would you want to drive on a busy highway or in a city where there were no traffic signs? Controlling the flow of traffic is vital if anyone is to get anywhere. Similarly, within sentences the flow of meaning is controlled by punctuation marks, the conventionally agreed-upon "traffic signals" of written communication. We do the same thing in spoken language by means of pitch, breath pauses, and emphasis. Directing the flow of ideas in writing is not really difficult, and a useful procedure when you're unsure of how to punctuate a sentence is to say it aloud as in normal conversation. Pay careful attention to where you pause naturally within the sentence—that's likely to be where you need some punctuation.

Many detailed guides to punctuation exist, and you may want to look at them if you have a lot of queries in this area. You will also find excellent advice on punctuation in the front or back sections of some standard college dictionaries. Meanwhile, the following suggestions are offered on the most common problems many engineers tend to have with punctuation.

COMMAS

Confusion sometimes exists about commas because frequently their use is optional. *Before we arrived at the meeting we had already decided how to vote* would be written with a comma after *meeting* by some and not by others, because some people tend to be heavy comma users while others go light on them. The question to ask is, Does adding or omitting a comma in a given sentence create noise? In general, if no possible confusion or strain results, the tendency in technical writing is to omit unessential commas.

Often, omitting a comma after introductory words or phrases in a sentence will cause your reader to be momentarily confused—as you would have been if there were no comma after the first word of this sentence. Here are further examples of missing commas causing noise:

After the construction workers finished eating rats emerged to look for the scraps.

In all the containers were in good condition considering the rough journey.

As you can see the efficiency peaks around 10–12%.

If an acoustic horn has a higher throat impedance within a certain frequency range it will act as a filter in that range which is undesirable.

Solution:

After the construction workers finished eating, rats emerged to look for the scraps.

In all, the containers were in good condition considering the rough journey.

> As you can see, the efficiency peaks around 10–12%.
>
> If an acoustic horn has a higher throat impedance within a certain frequency range, it will act as a filter in that range, which is undesirable.

Again, try saying these sentences aloud with their intended meanings. You'll find you put the comma—or pause—where it belongs almost without thinking.

One more point about commas: Most technical editors prefer what is called a *serial comma* when you list words or ideas within a sentence, as in *The serial comma has become practically mandatory in most scientific, technical, and legal writing.* You may have been told that the *and* joining the last two terms replaces the need for a comma, but this is not so in technical writing. See how the serial comma is useful in the following sentences by reading them aloud and noting how you need the pause before the *and*:

> Fresnel's equations determine the reflectance, transmittance, phase, and polarization of a light beam at any angle of incidence.
>
> Tomorrow's engineers will have to be able to manage information overload, communicate skillfully, and employ a computer as an extension of themselves.

A serial comma may also prevent confusion:

> Rathjens, Technobuild, Johnson and Turblex build the best turbines for our purposes.

Unless *Johnson and Turblex* is the name of one company, you will need a serial comma:

> Rathjens, Technobuild, Johnson, and Turblex build the best turbines for our purposes.

SEMICOLONS

Whether we like it or not, the semicolon seems to be disappearing from much engineering writing. Often it is replaced by a comma, which is an error according to traditional punctuation rules. More frequently, we simply use a period and start a new sentence, but then a psychological closeness might be lost. Look at these two examples:

> Your program is working well, however mine is a disaster.
>
> Take Professor Hixson's class. You'll find he's a great teacher.

The relationship between these statements could be better stressed by using a semicolon:

> Your program is working well; however, mine is a disaster.
>
> Take Professor Hixson's class; you'll find he's a great teacher.

Perhaps one reason we don't see many semicolons in engineering writing is that fewer and fewer people feel confident using them. Another possibility is that little noise results from using a comma or a period and new sentence, as in the previous examples. Note this pair of sentences:

> We wanted to finish the computer program yesterday; however, the network was down all afternoon.
>
> We wanted to finish the computer program yesterday, however, the network was down all afternoon.

Although the first sentence would be considered correct and the second wrong, you will find plenty of examples of the second punctuation around. The main problem in the second sentence is that a reader can't be sure at first whether *however* "belongs" to the first half of the sentence or the second. A semicolon after *yesterday* is really needed to make this clear. If you frequently use words like *however*, *therefore*, *namely*, *consequently*, and *accordingly* to link what could otherwise be two sentences, insert a semicolon before and a comma after them. You'll find this will add a shade of meaning that cannot be achieved otherwise.

Use semicolons to separate a series of short statements listed in a sentence if any one of the statements contains internal punctuation. The semicolon will then divide the larger elements:

> I suggest you choose one social science subject, such as psychology or philosophy; one natural science course, such as chemistry, physics, or biology; and one math class.
>
> The team is made up of Seth Deleery, vice-president of marketing; Natasha Kanapathy, director of research; Ruth Ustby, assistant director of training and human relations; and Leo Little, chief avionics engineer.

COLONS

Colons are used to separate the hour and minute in a time notation and to divide parts of book or article titles:

> This proposal is due on Monday morning at 8:30 sharp.
>
> One of the books recommended for the seminar is *The Limits of Safety: Organization, Accidents, and Nuclear Weapons.*

The most common use of the colon within a sentence, however, is to introduce an informal list:

> For the final exam you will need several items: a pencil, a calculator, and three sheets of graph paper.

You can also use a colon to introduce an illustration or example, as we did in the sentence leading into the last example. Note, however, that in both cases an independent clause—a statement that can stand by itself and have meaning—comes before the colon. You should *not* write the example sentence as

> For the final exam you will need: a pencil, a calculator, and three sheets of graph paper.

because what comes before the colon makes no sense by itself and the colon needlessly interrupts the flow of the sentence. Instead, write

> For the final exam you will need a pencil, a calculator, and three sheets of graph paper.

(Note how the same reasoning made us lead into the last two illustrations with no colon after the words "example sentence as" and "Instead, write.")

PARENTHESES

Use parentheses to set off facts or references in your writing—almost like a quick interjection in speech:

> Resistor R5 introduces feedback in the circuit (see Figure 5).
>
> This reference book (published in 1993) still contains useful information.

If what you place within parentheses is not a complete sentence, put any required comma or period outside the parentheses:

> Typical indoor levels of radon average 1.5 picocuries per liter (a measure of radioactivity per unit volume of air).
>
> Whenever I design a circuit (like this one), I determine the values of the components in advance.

If your parenthetical material forms a complete sentence, put the period inside the marks:

> I have already calculated the values of the resistors. (R1 is 10.5 KΩ, and R2 is 98 Ω.) The next step is to choose standard values.

Remember, it is best not to use parenthetical material too frequently because these marks force your readers to pause and are likely to distract them (if only for a brief moment—see what we mean?) from the main intent of your writing.

DASHES

A dash (often mistakenly referred to as a hyphen) will make a sentence seem more emphatic by calling attention to the words set aside or after it: *He was tall, handsome, rich—and stupid.* Because the dash is considered less formal than the other parenthetical punctuation marks (parentheses and commas), you should try to avoid it in very formal writing. If you overuse it, you are in danger of calling wolf too often, and your dashes will lose their effect. With this caution in mind, you may still find dashes helpful for the following purposes:

> Emphasis: Staying up all night to finish a lab project is not so terrible—once in a while.
>
> Summary: Reading all warnings, wearing safety glasses and hardhats, and avoiding hot materials—all these practices are crucial to sensible workshop procedure.
>
> Insertion: My opinion—whether you want to hear it or not—is that the drill does not meet the specifications promised by our supplier.

Notice that we're talking about the "em" dash here—the dash used between words that practically touches the letters at each end of it, and which we have used in this sentence. The "en" dash is shorter, more like a hyphen, and used when you cite ranges of numbers: *31–34; $350–400.* Most word processing programs allow you to choose whichever you need.

HYPHENS

Hyphens have been called the most underused punctuation marks in technical writing. Omitting them can sometimes create real noise, as when we read *coop* (an enclosure for poultry or rabbits) but discover that *co-op* was meant. On the other hand, a hyphen sometimes appears where it is unneeded, as in *re-design, sub-question,* or even *un-needed.*

Unfortunately, apart from the general rule that hyphens should be used to divide a word at the end of a line or to join pairs of words acting as a single descriptor—as in *The transistor is a twentieth-century invention*—there is no clear consensus on when to use them. You'll often have to decide for yourself with the help of a recent dictionary, but here are some suggestions:

- Don't hyphenate prefixes such as *pre-, re-, semi-,* and *sub-* unless leaving out a hyphen causes an eyesore or possible confusion. *Preconception* is fine, but *preexisting* needs a hyphen if only for looks. The same might be said of *antiinflationary, ultraadaptable,* or *reengineering.* You may have to distinguish between *recover* (regain) and *re-cover* (to put a new cover on) and the like at times. Again, a good dictionary will help.

- Don't hyphenate compound words before a noun when the first one ends in *ly.* For instance, *early warning system* needs no hyphen since it is clear that *early* modifies *warning,* not *system.* The same applies to *optimally achieved goals*, *highly sensitive cameras,* and similar constructions.

- Stay alert for sentences in which you can eliminate noise by adding one or more hyphens. As you can see, each of the following benefits from the insertion of hyphens at a critical point:

We used a 16 key keypad.
We used a 16-key keypad.

We knew Marienet made klystrons would be able to generate a 9.395 GHz microwave.
We knew Marienet-made klystrons would be able to generate a 9.395 GHz microwave.

The equation assumes a one dimensional plane wave propagation inside the horn.
The equation assumes a one-dimensional plane-wave propagation inside the horn.

Research showed the computer aided students improved their grades dramatically.
Research showed the computer-aided students improved their grades dramatically.

With really complex technical terms, you may have very little to go on regarding hyphens. For instance, how do you punctuate *direct axis transient open circuit time constant*? The best solution (*direct-axis transient open-circuit time constant*) may only be found in a technical dictionary or by observing what the common practice is among specialists in the field.

EXCLAMATION POINTS

The best advice on the exclamation point is to use it all you want in your novel or letters, but avoid it in professional writing except in the case of warnings (*DANGER: Sodium cyanide is extremely toxic!*). Because engineering documents seek to convey information, any excitement or triumph should be generated by the facts provided in the document rather than by a tagged-on marker. Occasionally an exclamation mark might even be interpreted by your readers as arrogant or sarcastic:

We soon found that the previous data was unsubstantiated!

After reading your report, I feel you might benefit from our on-site course in technical writing!

Punctuation error!

QUOTATION MARKS

Use quotation marks to set off direct quotations in your text, and put any needed period or comma within them, even if the quoted item is only one word. Although British publishers use different guidelines, the American practice is always to put commas and periods inside quotes, and semicolons and colons outside, as in the following examples:

The manager stressed to the whole group that the key word was "Preparedness."

"The correct answer is 18.2 Joules," he told me.

We had heard about the "Four-Star Marketing Plan," but no one remembered what it involved.

We left the game right after the band played "The Eyes of Texas"; it was too darn hot and humid to stay any longer.

Sometimes the question of where to put question marks with brief quotations arises. The solution is quite simple: If the question mark applies only to what is within the quotation marks, it goes inside the final quotation marks. No period follows after the quotation marks. If it applies to the whole sentence, it will go outside the final quotation marks:

Their manager bluntly asked, "Are we on schedule?"

What is the meaning of the term "antepenultimate"?

If you need to quote material that takes up more than two lines, set it off from your text by a space and indent it from both right and left margins. You might even use a slightly smaller font size, and should omit the quotation marks, as shown here:

According to the author, specifications should not be written by a single person:

The lead engineer delegates the writing of numerous sections to specialists, who may not be aware of the overall goals of the project, and may have parochial views about certain requirements. The lead engineer is faced with the difficult task of fitting all these pieces together, finding all the places where they may conflict, and adjusting them to be correct and consistent with each other [NAWCTSD Technical Report 93-022, p. 11].

The importance of consistency cannot be overstressed in the production of. . . .

SENTENCE SENSE

As an engineering writer, your aim is to convey information with a minimum of noise. Thus, the only important "rule" of grammar is to eliminate noise so that the readers of your document receive precisely the message you intend. In other words, your signal-to-noise ratio should be as high as possible. This section looks at the grammatical and stylistic areas where noise often seems to occur in engineering writing. Under the heading of "Two Latin Legacies," we also discuss two persistent but outmoded grammar rules you can safely forget.

CONNECTING SUBJECTS TO VERBS

It's unlikely you would write *The machines is broken* without quickly noticing a discrepancy between the subject (*machines*) and the verb (*is*). A problem can occur, however, when several words come between your subject and verb and you forget how you started the sentence. If you are writing in a hurry and leave no time for editing, you might produce something like this:

This <u>combination</u> of electrical components <u>constitute</u> a single-pole RC filter.

A 35 mm <u>film</u> of some high buildings <u>are</u> strongly recommended.

Only <u>one</u> of the pre-1925 high-rise structures <u>were</u> damaged in the quake.

Plural nouns that follow later (in these examples, *components, buildings, and structures*) can sometimes mislead us into relating the verb to them rather than to the earlier nouns (*combination, film, one*) to which they belong. This danger increases with the length of a sentence and the amount of information intervening between the true subject and verb of a sentence. A good style or grammar program on your word processor may help prevent this from happening, but it is just as well to be alert to the danger.

Sometimes a question about agreement arises in engineering writing regarding units of measurement. Do you write *Twelve ounces of adhesive were added* or *Twelve ounces of adhesive was added*? How about *12 grams of acid was spilled* or *12 grams of acid were spilled?* The answer is a matter of logic rather than grammar. Even though we're alluding to several ounces or grams here, we "see" them as one unit, and thus the singular verb is preferable. Little or no noise is created, however, if you slip up on this one.

Using *either/or* in a sentence occasionally makes us stop and think. Look at this sentence:

Either the old manual or the recent procedures (is/are?) acceptable.

Which verb should you use? Because a verb is normally controlled by the noun immediately before it, we would write *Either the old manual or the recent procedures* are *acceptable.* Following this practice, we could also write

Either the recent procedures or the old manual *is* acceptable.

It is best to follow the same rule with *neither/nor*. Thus, the following two sentences would be preferred:

Neither the engineers nor their supervisor *was* invited to the planning conference.

Neither the rudder nor the wings *were* badly damaged in the crash.

MODIFIERS

A modifier is a word or group of words whose function is to add meaning to other ideas in a sentence. If you say your company has bought a transceiver, you have certainly conveyed some meaning, but if you say *Our company has bought a TS 840 S transceiver with single sideband capabilities*, you add a lot of meaning to the word *transceiver* by adding some modifiers.

The danger lies in creating noise by misplacing the modifiers in a sentence. Such distortion can produce sentences that don't make sense or that make sense in the wrong way. Misplaced modifiers occur when a reader gets the wrong impression (or no impression) of who is doing what in a sentence. This is frequently because words like "I" or "we" or "the engineers" or some other subject have been omitted. Consider the following:

Jumping briskly into the saddle, the horse galloped across the prairie.

After testing the mechanism, the theory behind it was easily understood.

Once having completed needed modifications and adjustments, the equipment operated correctly and met all specifications.

If we look at these statements logically, we have a horse that rides, a theory that can test a mechanism, and equipment that modifies and adjusts. This is not likely to be what the writer meant. Revising the sentences might result in the following:

Jumping briskly into the saddle, the outlaw galloped across the prairie.

After testing the mechanism, we easily understood the theory.

Once we had completed needed modifications and adjustments, the equipment operated correctly and met all specifications.

Meanwhile, another problem can crop up if you place a modifier too far from the word or idea it modifies:

I was ordered to get there as soon as possible by fax.

By the age of four his father knew he would be an engineer.

It's not hard to remedy the lack of logic in these sentences and to avoid traveling by fax or having four-year-old fathers, but sometimes the meaning cannot be extracted, as in the following:

The tone-detector circuit was too unreliable to be used in our telephone answering device, which was built of analog devices.

The sentence is correct if the telephone answering device is made of analog devices, but much more likely the writer is concerned with the inaccuracies of an analog tone-detector circuit. This is easily fixed:

> The tone-detector circuit, which was built of analog devices, was too unreliable to be used in our telephone answering device.

UNCLEAR PRONOUNS

When you use a pronoun in your writing, it is commonly assumed that you are referring to whatever noun or nouns come just before it in the sentence. Thus, *The promotion was given to Vicky, who really deserved it,* is perfectly clear: The *who* refers to Vicky. Sometimes we get careless, though, especially with the pronouns *this* and *that*, their plurals, and *which* and *it*. Look at this example:

> We will study the terrain by soil analysis and computer simulation before reaching a decision on whether construction can take place here. This will also enable us to...

What does the *This* refer to in the second sentence—study, terrain, analysis, simulation, decision, or construction? According to accepted usage, it should be *construction* because it's the last noun before the pronoun *This*, but that's unlikely to be what the writer meant. The meaning would be much clearer if the second sentence read something like this:

> This study will also enable us to...

Let's look at another example:

> Ambiguous: Back in 1954, three researchers made a series of discoveries about the unknown sources of Barbour's early notebooks. These prompted them to further investigate . . .
>
> Clearer: These discoveries prompted the three to further investigate . . .

PARALLELISM

Parallelism refers to the need for items in a list to share the same grammatical structure. Faulty parallelism creates noise because it violates a sense of logical consistency. Rather than tell someone you *like to jog, wrestling, and play the fiddle,* you would probably say you *like to jog, wrestle, and play the fiddle,* or that you enjoy *jogging, wrestling, and playing the fiddle.* But in longer sentences there is a danger of losing control of this logic.

After a lot of discussion the team concluded that their alternatives were to call in a consultant, thus increasing the cost of the project, or having three more engineers reassigned to the team.

Note how this sentence reads as if the team's alternatives are (1) to call in a consultant, and (2) having more engineers reassigned—two unparallel statements that can grate on our sense of logical flow. The sentence can be rewritten to state that the alternatives were *to call in a consultant . . . or to have three more engineers reassigned.*

See if you can recognize the lack of parallelism in this sentence:

The back-up system should be efficient, should meet safety specifications, and have complete reliability.

To make this statement parallel, think of the list embedded in it. We are told that the back-up system

1. should be efficient
2. should meet safety specs
3. have complete reliability

To be consistent, the sentence needs one more *should*—or one less:

The back-up system should be efficient, should meet safety specifications, and should be completely reliable.

The back-up system should be efficient, meet safety specifications, and be completely reliable.

This might seem like a rather fine point, but because a lack of parallelism can often cause a reader to pause, if only subconsciously, it qualifies as noise when it occurs in a sentence. Keeping parallel structure is even more important when you construct lists, as we pointed out in Chapter 2 in the section "Use Lists for Some Information."

FRAGMENTS

Sentence fragments are partial statements that create noise because they convey an incomplete unit of information. Here's an example:

> She decided to major in petroleum engineering. Even though it would take five years.

The first sentence makes sense by itself. Try saying the second statement alone, as an independent exclamation, and your listeners will be lost.

We must admit, however, that in everyday speech and popular journalism you will find plenty of fragments that seem to cause little or no noise. Look at this example:

> Nearly 60% of U.S. households had VCRs by the end of the 1980s. In spite of the microwave oven being the most popular appliance of the decade.

We know what the writer means here, but strictly speaking the second statement is a fragment because it could not stand alone and make sense. The words *In spite of* indicate a contrastive relationship that is clear only in the context of the first statement. It would be more efficient to write

> In spite of the microwave oven being the most popular appliance of the 1980s, nearly 60% of U.S. households had VCRs by the end of the decade.

In your formal engineering writing you would do well to avoid incomplete sentences. They can usually be quite easily remedied, as you can see. Here's another example:

> Fragment: *Delays in the October shipments have occurred. Due to the strike.*
>
> Complete: *Delays in the October shipments have occurred due to the strike.*
>
> Better: *The strike has delayed the October 6 shipments.*
>
> or
>
> *The October 6 shipments have been delayed by the strike.*

ACTIVE OR PASSIVE VOICE?

As indicated in the last pair of sentences, we can use two distinct "voices" in English sentences. The active voice directly states that someone does something, as in *The engineer wrote the report.* The passive voice turns it around to *The report was written by the engineer*. Thus, the active voice emphasizes the performer of the action—the engineer, in our example—whereas the passive emphasizes the recipient of the action, the report.

Many engineering and scientific writers are told to use the passive voice, that is, to leave themselves out of their writing. They might write *It was ascertained that. . .* rather than *We made sure that. . .* , or *The deadline was met* rather than *We met the deadline*. Chances are management would rather tell you *It has been decided to terminate your employment* than *We have decided to fire you*. (Perhaps such hedging is necessary at times because it helps conceal responsibility and gives us no one to blame!)

The passive voice is certainly appropriate when writing up your research or describing a process, for example. There are plenty of instances where you don't want the "doer" to get in the way of your description. Also, it's logical to use the passive if the doer of an action is unknown or unimportant, or if what is being done is simply more important than who did it:

> Electricity was discovered thousands of years ago.
>
> The bridge was torn down in 1992.
>
> The contaminated material is then taken to a safe environment.

Sometimes the passive will give variety to your writing, even if your inclination is to write predominantly in the active voice:

> Computer experts claim that general-purpose processors have unpredictable execution times due to their use of complex architectural features. This conjecture has now been tested by our group and we have found that the architecture really induces little or no unpredictability. Moreover, data gained from our study show how the execution times can be predicted. It was also found that . . .

In spite of the passive's usefulness, however, the natural form of the English sentence is usually the active voice. This form generally tends to be the more efficient one. Look at the following pairs:

> *Control of the flow is provided by a DJ-12 valve.*
> A DJ-12 valve controls the flow.
>
> *A system for delineating these factors is shown in Figure 5.*
> Figure 5 shows a system for delineating these factors.
>
> *By switching off the motor when it started to vibrate and looking at the tachometer, the resonant frequency was determined.*
> We determined the resonant frequency by switching off the motor when it started to vibrate and looking at the tachometer.

The passive can become especially burdensome in procedures or instructions:

The button is pressed twice.

vs.

Press the button twice.

Previously entered data in the DataBase is eliminated by the Edit menu being opened and Select All being chosen.

vs.

Eliminate previously entered data in the DataBase by opening the Edit menu and choosing Select All.

Nowadays engineering writers are tending to get away from the rigid use of the passive as they realize there is a lot to be said for using the active voice. Sentences become more vigorous, direct, and efficient in the active form, and by showing that a *person* is involved in the work, you are doing no more than admitting reality. Also, the active voice gives credit where credit is due. If we read in a progress report that *several references were checked out from the library and 25 pages of notes were taken,* are we as impressed by the energy expended as when we read *I checked out several books from the library and took 25 pages of notes?*

One danger of avoiding the active voice is that we can end up saying some pretty awkward things:

Hurrying to complete the project, several wires got soldered together incorrectly.

The supervisor was seen by us, and we were ignored by her.

My darling . . . you are really loved by me. Am I loved by you?

Perhaps the best policy is to use the active voice in your technical (and romantic) writing if it seems the most natural and efficient way to express yourself, assuming there is no company policy against its use. Don't hesitate to write in the passive, however, if the circumstances seem to call for it or if the specifications for the document you are writing require it.

SEXIST LANGUAGE

Gender, or sex, is now only indicated in English by *she/he, his/hers, her/him,* and by a small group of words describing activities formerly pursued by one sex or the other, such as *mailman, stewardess, chairman,* or *seamstress.* Now, of course, men might bring the drinks on an airplane and women might deliver the mail, not to mention take an equal place in the engineering workplace. Given this situation, it is unnecessarily restrictive—and to some people offensive—to use gender-specific terms in writing and speech unless there is good reason to do so. The following pairs show how easy it is to reword your sentences or paragraphs to include everyone they should:

Restrictive: *Every engineer should be at his workstation by 9 a.m.*
Inclusive: *Every engineer should be at his or her workstation by 9 a.m.*

        or (preferred because less wordy):

        Engineers should be at their workstations by 9 a.m.

Restrictive: *An employee can expect a lot of challenges during his career here.*

Inclusive: *Employees can expect a lot of challenges during their careers here.*

Restrictive: *Every technician must wear safety glasses when he enters the work area.*

Inclusive: *Technicians must wear safety glasses when entering the work area.*

Most nouns indicating gender in English have already been modified to be inclusive. A recent dictionary can guide you here. One title that still sneaks through, however, especially in organizations traditionally dominated by males, is *chairman*. If the "chairman" is female, is she the *chairwoman* or *chairperson*? Both are acceptable, but it's probably simpler to refer to anyone in such a position as the *chair:*

Sarah is chair of the new committee on marketing strategy.

       or

Sarah is chairing the new committee on marketing strategy.

TWO LATIN LEGACIES

A few grammar rules impressed upon us in the past really do not hold up under careful linguistic or logical inspection. They were based on how Latin works, rather than English. To put it another way, noise rarely occurs when these rules are ignored. Here are the two main ones, together with comments and a caution.

"Never End a Sentence with a Preposition." In reality, a preposition is often the best word to end a sentence with. (A purist might claim we should have just written *the best word with which to end a sentence.*) When an editor criticized Sir Winston Churchill for doing so, Churchill responded with "Young man, this is the kind of nonsense up with which I will not put!" After all, did you find any noise in the opening sentence of this paragraph? Efficient writing sometimes dictates that we end a sentence with a preposition. Compare the following pairs. You can see that in each case the first sample, ending with a preposition, flows better and is more natural:

> That's a problem that we will really have to work on.
>
> That's a problem on which we will really have to work.
>
> We must make sure we can find some engineering consultants we can really count on.
>
> We must make sure we can find some engineering consultants on whom we can really count.

"Never Split an Infinitive."   An infinitive is the form of a verb that combines with the word *to*, as in *to go*, *to work*, or *to think*. Confident writers have dared *to deliberately split* the infinitive whenever doing so was in the best interests of clear writing. Certain TV space adventurers have been daring *to boldly go* where the rest of us can't for a long time now, and an electrician may find it necessary (and safer) *to entirely separate* the wires in a power line sometimes. But don't overload a split infinitive. If you put too much material between your *to* and the rest of the verb, noise or even nonsense might result:

> The team has been unable to, except for the lead engineer and one technician who is on temporary assignment with us, master the new program.

Rewrite this as

> Except for the lead engineer and one technician on temporary assignment with us, the team has been unable to master the new program.
>
>   or
>
> The team has been unable to master the new program—with the exception of the lead engineer and one technician who is on temporary assignment with us.

TRANSITIONS

Transitional words and phrases are signposts that show a reader the way your thinking is going. They help connect ideas, distinguish conditions or exceptions, or point out new directions of thought. Simple words like *therefore*, *thus*, *similarly*, and *unfortunately* eliminate ambiguity by helping a reader interpret your information. So if you neglect transitions in your writing, you may create noise, because your reader might miss some important connection. Look at these two sentences:

> The group's long-range plans for the S-34B project have been extended. The completion date for the project is as originally planned.

Both sentences are grammatically correct and contain important facts, but can the reader tell how these facts are related? Now notice how the next three illustrations indicate relationships the first example does not:

> The group's long-range plans for the S-34B project have been extended. Nevertheless, the completion date for the project is as originally planned.
>
> The group's long-range plans for the S-34B project have been extended. Unfortunately, the completion date for the project is as originally planned.
>
> Even though the group's long-range plans for the S-34B project have been extended, the completion date for the project is as originally planned.

Although facts are important, it is often the *relationships* between the facts that create the whole picture. Thus, you should make your transitions and connections as strong as possible. Here are some examples.

> To indicate a sequence: *before . . . later, first . . . second, in addition, additionally, then, next, finally*
>
> *Before the project got under way we felt we could never meet the deadline. Later, it became clear there was a realistic chance of doing so.*
>
> To indicate contrast: *but, however, yet, still, nevertheless, although, on the contrary, in contrast, on the other hand*
>
> *The GX-40 vehicle scored over 96% in initial dependability testing; nevertheless, the design was scrapped.*
>
> To indicate cause and effect: *consequently, therefore, so, thus, hence*
>
> *This company has had to downsize lately. Consequently, many of our staff are looking for other positions.*
>
> To indicate elaboration: *further, furthermore, for example, moreover, in fact, indeed, certainly, besides*
>
> *The automotive airbag has proved to be a major factor in driver survival. Moreover, the bag has generated considerable profits for its producers.*

SENTENCE LENGTH

When dealing with highly technical subjects, you should rarely write sentences over 20 words long. Technical material can be difficult enough to follow without being presented in lengthy, complex sentences. This difficulty increases if your audience is less familiar with your field than you are. Even nontechnical ideas are hard to grasp in an unnecessarily long-winded sentence:

> We finally had a long discussion with the R&D staff but were not able to convince them that they should commit to a specific date for implementation of the design, but instead they responded with a proposal to extend the project, which would result in a lot more work for all of us and a considerable loss of profits for the company.

Nobody wants to be left breathless at the end of a mile-long sentence. If you find your sentences tend to be lengthy, look for ways to break them into two or more separate ones. The readability of your prose will be determined partly by the length of your sentences. On the other hand, too many short sentences may leave your readers feeling like first graders:

> The Kw766XTR is a low-profile desktop scanner. It has outstanding performance. It offers a frequency range of 29–54 and 108–174 MHz. It includes 50 memory channels. The design is sleek. Individual channels can be locked out. They can also be delayed.

Try to vary your style and avoid both lengthy and abrupt sentences. Very short sentences used sparingly, however, can be effective in helping you reinforce a point. Remember this.

TECHNICAL USAGE

USELESS JARGON

In its negative sense, *jargon* is pure noise because it refers to unintelligible speech or writing. The word derives from a French verb meaning the twittering of birds, and has a lot in common with "gobbledygook," first used to compare the speech of Washington politicians to the gobbling of Texas turkeys. High-tech jargon is sometimes known as techno-babble or scispeak. Some people seem to like to sprinkle their writing liberally with such impressive-sounding phrases as *integrated logistical programming, differential heterodyne emission,* or *functional cognitive parameters.*

Unfortunately, unless these words hold a precise meaning for both writer and reader, no communication takes place.

Techno-babble is so common that with tongue in cheek we have created an "electrotechnophrase generator" in Figure 3-1 to help addicts satisfy their habit. Select any three-digit number and read off the corresponding words from the chart; for example, 2-8-3 generates *differential heterodyne emission.* Readers may have no idea what you mean, but they should be impressed—or afraid to ask for a meaning.

USEFUL JARGON

In another sense, however, jargon is the necessary technical terminology used in specialized fields. A chemist might use the term *deoxyribose* around a group of peers without feeling a need to explain it, just as a geologist could talk about the Paleozoic era or Devonian period with other geologists. Computer engineers can safely refer to bytes, bauds, and packet switching—among themselves. Communication between experts would be ponderous, if not impossible, if they had no specialized jargon. Moreover, each year technical language increases greatly as scientific knowledge increases; thousands of technical terms used today were unknown just a few years ago.

Sometimes you will find that common words take on new meanings when used by experts. *Charge, conductor, mole,* and *mud* are just four examples. Typesetters mean something quite different than most of us would when they refer to *widows, orphans,* and *leading.* As engineers, you know and use all sorts of technical jargon. Some you share with practically all engineers, some with those in the same general field of engineering as you—such as chemical, civil, or aerospace—and some you would use only among peers in highly specialized fields such as celestial mechanics or software engineering.

	Column 1	*Column 2*	*Column 3*
0.	voltaic	integrated	simulation
1.	Sholokhov's	semiconductor	algorithm
2.	differential	Yagi	attenuator
3.	Fourier	scaled	emission
4.	transient	Q-factor	diode
5.	virtual	tracking	parameters
6.	phasor	diffusion	network
7.	compound	Doppler	gate
8.	thermal	heterodyne	transducer
9.	Gaussian	coaxial	magnetron

Figure 3-1   The electrotechnophrase generator (courtesy of ECE students at the University of Texas at Austin).

There is only one way to avoid noise when using technical terminology: *Know your audience.* Make certain you are writing or speaking at their level of comprehension, because if you're above their heads you will be wasting your time and theirs. Explain terms whenever necessary; don't risk confusing readers or completely losing them because they don't know what you are talking about. Definitions within your text, examples, analogies, or a good glossary are all useful tools for the technical writer who must frequently communicate with less technically inclined audiences. These specific tools are discussed more fully in other sections of this book.

ABBREVIATIONS

Abbreviations are necessary in technical communication for the same reason valid technical jargon is: They refer to concepts that would take a great deal of time to spell out fully. It would be time-consuming and boring for a computer expert to read *computer-aided design/computer-aided manufacturing* several times (or hear it in a talk) when *CAD/CAM* would do. However, you will create a lot of noise in your writing if you use abbreviations your readers don't understand. Always spell abbreviations out the first time you use them unless you know this would insult the intelligence of your audience:

> Then it goes into the ROM (read-only memory).
>
> To understand our billing process, you first need to know what a British thermal unit (BTU) is.

Once you have defined an abbreviation, you can normally expect your reader to remember it. The exception to this would be if you are using some highly complicated or unusual abbreviations throughout your document, in which case you may need to remind readers more than once what the abbreviations stand for, or provide a glossary they can refer to.

Initialisms and Acronyms.   Abbreviations can be subdivided into initialisms and acronyms. *Initialisms* (sometimes called *initializations*) are formed by taking the first letters from each word of an expression and pronouncing them as initials: *GPA, IBM, LED, UHF. Acronyms* are also created from the first letters or sounds of several words, but are pronounced as words: *AIDS, FORTRAN, NAFTA, NASA, RAM, ROM.* Some acronyms become so well known that they are thought of as ordinary words and written in lowercase: *bit, laser, pixel, radar, scuba, sonar.*

Don't be surprised if you find a list of both initialisms and acronyms lumped under the title "Acronyms." Many engineering writers no longer observe the distinction between the two, and call any abbreviation an acronym. You probably shouldn't make an issue of it, especially if the writer is your superior.

Two usage pointers are as follows:

1. Use the correct form of *a/an* before an initialism. No matter what the first letter is, if it is pronounced with an initial vowel sound (for example, the letter *M* is pronounced "em") write *an* before it:

   an MTCR (missile technology control regime)
   an LED readout
   an SRU pin
   an ultrasonic frequency (but *a UHF receiver*)

   Some abbreviations might fool you. Consider LEM (lunar excursion module), for example. If the custom is to pronounce it as an initialism, L-E-M, then you will have *an* LEM. If it is normally considered an acronym (as one word), you will have *a* LEM.

2. Form the plural of acronyms and initializations by adding a lowercase *s*. Only put an apostrophe between the abbreviation and the *s* if you are indicating a possessive form:

   We ordered three CRTs.
   We weren't satisfied with the last CD-ROM's performance.
       or
   We weren't satisfied with the performance on the last CD-ROM.

NUMBERS

Engineering means working with numbers a great deal. Frequently, this is where a lot of written noise occurs due to typos, incorrect or inexact numbers, and inconsistencies. Obviously, you can avoid serious noise by making certain any number you write is accurate. You should also give numbers to the necessary degree of precision: Know whether 54.18543 is needed in your report or whether 54.2 will do. Avoid noise from inconsistent use of numbers by following these guidelines:

1. Numbers are expressed as words (twelve) or numerals (12). Cardinal numbers are *one, two, three*, etc. Ordinal numbers are *first, second, third*, etc. Although custom varies, it's a good idea to write the cardinal numbers from one to ten as words and all other numbers as figures.

two transistors	232 stainless steel bolts
three linear actuators	12 capacitors

   However, when more than one number appears in a sentence, write them all the same:

   The IPET has 4000 members and 134 chapters in 6 regions.

   Also, use numerals rather than words when citing time, money, or measurements:

1 a.m.	$5.48	12.4 m	8 ft

2. Spell out ordinal numbers only if they are single words. Write the rest as numerals plus the last two letters of the ordinal:

> second harmonic   21st element   fourteenth attempt   73rd cycle

3. If a number begins a sentence, it's a good idea to spell it out regardless of any other rule.

> Thirty-two computers were manufactured today.

To avoid writing out a large number at the beginning of a sentence, rewrite the sentence so it doesn't begin with a number:

> Last year, 5198 engines were manufactured in this division.
>
> or
>
> *This division manufactured 5198 engines last year.*

Note   You may sometimes see very large numbers written with spaces where you expect commas. Thus 10,354,978 might appear as 10 354 978. This style avoids any possible confusion with the practice in some countries of using commas as decimal markers. Decide which method you want to use based on your company's preference and your audience.

---

4. Form the plural of a numeral by adding an *s*, with no apostrophe:

> 80s   1920s

Make a written number plural by adding *s* or *es*, or by dropping the *y* and adding *ies:*

> nines   sixes   fours   nineties

5. Place a zero before the decimal point for numbers less than one. Omit all trailing zeros unless they are needed to indicate precision.

> 0.345 cm   12.00 ft   0.5 A   19.40 tons

6. Write fractions as numerals when they are joined by a whole number. Connect the whole number and the fraction by a hyphen:

> 2-1/2 liters   32-2/3 km

7. Time can be written out when not followed by a.m. or p.m., but you will normally need to be more precise than this. Use numerals to express time in hours and minutes when followed by a.m. and p.m. or when recording data. Universal Time (UTC, from the French for *universal coordinated time*) uses the 24-hour clock.

> ten o'clock   10:41 a.m.   8:45 p.m.
> 4 hours 36 minutes   12 seconds   23:41 (= 11:41 p.m.)

8. When expressing very large or small numbers, use scientific notation. Some numbers are easily read when expressed in either standard or scientific form. Choose the best format and be consistent:

> 0.0538 m or 5.38×10^{-2} m 8.32×10^{-21} m/s or   367 345 199 m/s

UNITS OF MEASUREMENT

Although the public in the United States is still not committed to the metric system, you will find that in general the engineering profession is. Two versions of the metric system exist, but the more modern one, the SI (from French *système international*), is preferred. The vital rule is to be consistent. Don't mix English and metric units unless you are forced to. Be sure to use the commonly accepted abbreviation or symbol for a unit if you do not write out the complete word, and leave a space between the numeral and the unit.

70 ns	100 dB
12 V	34.62 m
23 e/cm^3	6 Wb/m^2

Many people, including technically trained ones, still think in standard or English units of measurement, so sometimes you may find it advisable to give both referents in your writing. As with many other editorial matters, you can only make this decision after thinking of your readers' needs. When it might be advisable to add "explanatory" units, as with a mixed audience, do so by writing them in parentheses after the primary units:

212°F (100°C)	5.08 cm (2 in.)

Make sure you use the correct symbol when referring to units of measurement, and remember that similar symbols may stand for more than one thing. A great deal of noise (or disaster) could result if you confused the following, for example:

°C (degrees Celsius)	C (coulomb—unit of electric charge)
g (gram)	G (gauss—measure of magnetic induction)
m (thousandth)	M (million)
n (nano-)	N newton
s (second—as in time)	S (siemens—unit of conductance)

Units of measurement derived from a person's name usually are not capitalized, even if the abbreviation for the unit is. Note also that although the name can take a plural form, an *s* is not added to the abbreviation to make it plural:

amperes A	farads F	henrys H	webers Wb
kelvins K	teslas T	volts	

When working with very large or very small units of measurement, you will need to be familiar with the designated SI expressions and prefixes:

Factor	Prefix	Symbol
10^{18}	exa-	E
10^{15}	peta-	P
10^{12}	tera-	T
10^9	giga-	G
10^6	mega-	M
10^3	kilo-	k
10^2	hecto-	h
10^1	deka-	da
10^{-1}	deci-	d
10^{-2}	centi-	c
10^{-3}	milli-	m
10^{-6}	micro-	μ
10^{-9}	nano-	n
10^{-12}	pico-	p
10^{-15}	femto-	f
10^{-18}	atto-	a

A recent dictionary of scientific terms will guide you if you are unsure of the correct spellings or symbols of the units you are using. There is no point using them in your writing, however, if you or your audience don't know what they mean. Symbols and abbreviations are indispensable to an engineer, but use them sparingly when writing for an audience other than your peers. You may sometimes need to define the ones you use, either in your text parenthetically (a brief explanation in parentheses following the term or symbol, like this) or with annotations, as in the following example:

$$P = I\,E \qquad\qquad (1)$$

where

P = power, measured in watts

E = EMF (electromotive force) in volts

I = current in amperes

EQUATIONS

It would be hard to do much engineering without equations. They can communicate ideas far more efficiently than words can at times—consider the ideas represented by $E = mc^2$, for example. However, formulas and equations slow down your reader, so use them only when necessary and when certain your audience can follow them.

Many word processing programs now make it easy to write equations in text, but if you have to write them in longhand, do so with care to ensure both accuracy and legibility. An illegible or ambiguous equation is hardly going to communicate data effectively, and an error in an equation could be fatal. In other words, make sure your equations are noise free.

You should normally center equations on your page and number them sequentially in parentheses to the right for reference. Leave a space between your text and any equation, and between lines of equations. Also, space on both sides of operators such as =, +, or −, as shown in the following equations. If you have more than one equation in your document, try to keep the equal signs and reference numbers parallel throughout (see Figure 3-2).

Eventually you may have to incorporate multiline equations into your technical papers and reports, where they will read (and should be punctuated) just like sentences. As Figure 3-3 illustrates, no material is too complex to be presented clearly in a flowing, natural manner. Punctuation, transitions, accurate grammar, and mechanics are all indispensable tools for conveying highly technical information with a minimum of noise.

EDIT, EDIT, EDIT

If you look at the early handwritten drafts of some of the greatest writers' works, you'll see alterations, additions, deletions, and other squiggles that indicate how much revision went into the draft before it became a finished work. We could all produce better written documents if we always

1. *Had* the time to edit our work carefully
2. *Took* the trouble to edit our work carefully

$$F(x) = \int \log x \, dx \qquad (1)$$

$$H(s)\,(xv_2) = X(s)/Y(s) \qquad (2)$$

Figure 3-2   An efficient display of multiple equations.

The total harmonic distortion (THD) of voltage at any bus k is defined as

$$\text{THD}_k = \frac{\sqrt{\sum_{h=2}^{H}\left|V_k^h\right|^2}}{\left|V_k^1\right|}.$$

(3)

THD can be incorporated into the minimization procedure in (2) by considering a network function that equals the sum of squared THD_k's, or

$$f(I_m) = \sum_{k=1}^{K}\left(\text{THD}_k\right)^2 = \sum_{k=1}^{K}\left[\frac{\sqrt{\sum_{h=2}^{H}\left|V_k^h\right|^2}}{\left|V_k^1\right|}\right]^2,$$

$$= \sum_{h=2}^{H}\sum_{k=1}^{K}\frac{1}{\left|V_k^1\right|^2}\left|V_k^h\right|^2.$$

(4)

Note that (4) is identical to (2) when $y(h) = 1$ for $h = 2, 3, 4, \cdots, H$, and when

$$b(k) = \frac{1}{\left|V_k^1\right|^2}, \quad k = 1, 2, 3, \cdots, K.$$

(5)

Since the fundamental frequency voltages are approximately 1.0 pu, the objective function of (4) is a close approximation to that of (1).

Figure 3-3   An example of clear multiline equations.

For an engineer, time is frequently going to be a problem. You can't always find time for a leisurely edit of your work. However, you would be ill-advised to send a first draft of anything of importance to your readers. A quick email note to a friend about lunch isn't worth much concern, but anything more than this, especially if it's going beyond your immediate colleagues, needs at least to be looked over briefly with an editorial eye. How much time you invest in editing should be in direct proportion to the importance of the document. Use all the assistance your word processor will give you, including any spelling, grammar, or readability programs you may have, but don't follow their suggestions blindly. *You* have to be the final arbiter on the clarity and effectiveness of your work—*your* name will be on the document, not your word processor's manufacturer.

COLLABORATIVE PROOFREADING

There is nothing wrong with having a colleague, friend, or spouse look over your writing before you submit it to its intended audience. Two heads are usually better than one for discovering flaws in a piece of writing, and you are no longer in a freshman English class where such help might be considered plagiarism. In industry, experts often cooperate in writing technical reports, proposals, and other documents in the same way that they work together on engineering projects. In fact, most lengthy documents are produced by team effort, where different team members use their particular strengths to ensure that the document is the best it possibly can be.

Collaborative editing can involve something as simple as asking a friend for his or her opinion of the organization, clarity, and mechanics of your work, and using those comments to improve your writing where necessary. The more skilled and frank your friend is, the better. With a long document, however, collaborative editing can be done by having different team members check the document at different levels, which is usually better than having everyone searching for whatever they can find at all three levels at once. See the section in Chapter 2 entitled "Share the Load: Write as a Team" for more specifics on how to do this.

EXERCISES

1. Review some of your own recent writing for problems with spelling, punctuation, or any of the items listed in this chapter under "Sentence Sense." Did you create any noise in your documents by not following these guidelines? How could you use the guidelines as a quality-control tool when writing in the future?

2. Find what you feel is a good example of technical writing in any field. Analyze it carefully. What makes it effective, noise-free writing? List and give examples of the ways in which the writer has carefully observed many of the guidelines given in this chapter.

3. Look at an article in any professional journal and determine who its assumed audience is. Then investigate how the author uses technical terminology. Is it appropriate for the audience? Are explanations or definitions given where they seem called for? Do you find any examples of unnecessary technical jargon? How might such jargon have been avoided?

4. Check a number of reports or articles in technical journals that contain abbreviations, numbers, units of measurement, and equations. Are the authors consistent in the way they write these? Does the way these items are written vary from one report to the next, or from one journal to another? In the case of journals, is any information provided on how such things are to be written? Is there any indication in the journal that a style guide is available for writers who might wish to contribute articles?

BIBLIOGRAPHY

The Chicago Manual of Style, 14th ed. Chicago: The University of Chicago Press, 1993.

Pearsall, Thomas E. *The Elements of Technical Writing*, 2nd ed. Needham Heights, MA: Allyn & Bacon, 2001.

Rude, Carolyn, and Sam Dragga. *Technical Editing*, 3rd ed. New York: Longman, 2002.

Strunk, William, and E. B. White. *The Elements of Style*, 4th ed. New York: Allyn & Bacon, 2000.

Wired Style: Principles of English Usage in the Digital Age, rev. ed. Constance Hale, ed. San Francisco: HardWired, 1999.

4

WRITING LETTERS, MEMORANDA, AND EMAIL

A recent . . . survey found that the average employee spends nearly an hour a day handling e-mail chores. For managers, e-mail tasks usurp closer to two hours each day. It's no wonder people are complaining about e-mail fatigue.

Paul McFedries, "The Age of High (Tech) Anxiety,"
IEEE Spectrum (June 2003), p. 56

As a professional engineer, you should become familiar with the style, format, and content for business communications. (With contemporary "downsized" organizations, it may be a while before you rise high enough in the firm that you can rely on a secretary.) This chapter explores strategies for deciding which medium of communication to use and then moves on to discuss format, style, and content for business letters, memoranda, and email. The chapter concludes with writing-style issues that apply to any of the media described here.

WHICH TO USE?

Working professionals have at their disposal a variety of communication media. If you have a question for someone in your building, you can run down the hall and ask in person; if it's for someone within your organization but at a different location, you can write a memo; if it's for someone external to your organization, you can write a business letter; if it's urgent or informal, you can make a telephone call or send email.

PHONE OR PAPER?

The decision whether to use telephone or face-to-face communication as opposed to written communication is fairly obvious. In telephone or face-to-face communication, these are the issues:

- *Permanent record.* There is no record of what transpires in your phone conversation.
- *Availability of the recipient.* Recipients of the communication may not be in their offices or at their desks, forcing you to play "telephone tag."
- *Attitude of recipients.* Recipients may not take the phone or in-person communication as seriously as they would if it were in writing.
- *Purpose, length, and complexity of the topic.* Some topics are just too much for a conversation. For example, you can't present details of product specifications or a proposal over the phone.

EMAIL OR PAPER?

The decision gets harder when you choose between email and print. If you use email, you may wonder why you should bother with phone calls, business letters, or memos at all. Email eliminates the bother of stamps, envelopes, and mailboxes—not to mention the delay in delivery and response. Unlike telephone communication, email doesn't require its recipients to be in the right place at the right time—they can read it when they are ready. And, unlike telephone communication, email constitutes a record of the communication, although viewed by some as unofficial. However, print remains the preferable medium in certain instances, and sometimes the only communication medium. Here are the issues to consider:

- *Recipients.* Obviously, if some of the recipients don't have email or can't access email in certain circumstances, printed letters or memos are necessary.
- *Need for reply or forwarding.* If the letter or memo contains pages that the recipient must fill out and send, hardcopy print may be preferable.
- *Security issues.* As Ed Krol pointed out in *The Whole Internet User's Guide and Catalog* in the early days of the Internet, you can assume that any email you send has a chance of being seen by anyone in the world. Think twice about sending confidential information (new product specifications, confidential data about a project, or sensitive information about a colleague) by email.
- *In-person discussion of the memo.* If the message must be used in a face-to-face situation, print may be preferable. If everybody must print the memo for the meeting, you might as well send it in print and thus eliminate a potential snag.
- *Importance or length of the information.* For some, email lacks the feeling of settled, established information. It seems light, ephemeral—not a medium for

serious business. Some people are less likely to take an electronic message seriously than they are a hardcopy memorandum or letter.

- *In-your-face factor.* For some, a printed memo sitting on their desk just cannot be avoided. Of course, that depends—for some professionals, hard-copy mail is more inconvenient than email. Ultimately, you have to base your decision on which medium your colleagues are most in the habit of using.

LETTER OR MEMO?

Memoranda are written communications that stay within an organization (a business firm or a government agency, for example). Business letters are written communications to recipients who are external to the organization of the sender. Of course, some internal communications are in the form of business letters—for example, those letters that the CEO sends out once or twice a year to all employees.

BUSINESS LETTERS

As suggested earlier, the common business letter (printed on real paper!) is not dead. Face-to-face, telephone, and email communications are just not right for certain kinds of correspondence. Use a hardcopy letter when you want to make sure that the recipient receives it and takes it seriously, when you want the recipient to study it at length and act appropriately upon it, when the communication is long and packed with information, or when you want a permanent record of the communication. Use the following design suggestions for business letters—professional communications external to your organization.

STANDARD COMPONENTS OF BUSINESS LETTERS

The following describes standard components for business letters, most of which are illustrated in Figure 4-1. Of course, not all these components occur in any individual letter.

- *Company or personal logo.* If you use company stationery, begin your letter about an inch below the logo. Don't use logo stationery on following pages; use the matching stationery without the logo. If you are an independent consultant, design your own logo! Create a logo with a larger, fancy type style, maybe some combination of bold and italics, and maybe horizontal lines above or below your name, title, and address. (See the examples in Figures 6-1 and 10-9.)
- *Heading.* The heading contains the sender's address and the date. If you're using letterhead stationery, only the date is needed.

- *Inside address.* This portion includes the name, title, company, and full address of the recipient of the letter. Make this the same as it appears on the envelope. This element becomes important when secretarial staff discards the envelope.

- *Subject line.* Some business-letter styles make use of a subject line, the same kind that you see in memoranda. This element announces the topic, purpose, or both of the letter—for example, "Request for copyright status on the XI1 documentation" or "In response to your request for copyright status." (See Figure 4-4 for an example.)

- *Salutation.* This is the "Dear Sir" element. In contexts where no obvious recipient exists or where the recipient does not matter, omit the salutation. If you must include a salutation but don't have a specific name, call the recipient's organization (ask also for title and department name) or create a depart-

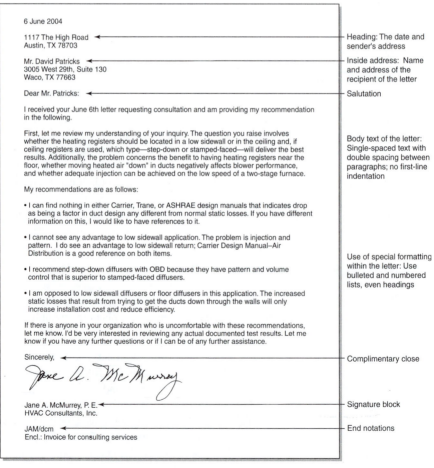

Figure 4-1   Standard business letter format—block letter.

ment or group name that is reasonably close. For example, use "Dear Recruitment Officers:". If all else fails, you can use the infamous "To Whom It May Concern:". Notice that the salutation for business letters is punctuated with a colon. (A comma implies a friendly, nonbusiness communication.)

- *Body of the letter.* The body begins just after the salutation and continues until the complimentary close. Text is single spaced; first lines of paragraphs are not indented; and double spacing is used between paragraphs. (For writing strategies and style to use in the body, see "Writing Styles for Business Correspondence," later in this chapter.)

- *Complimentary close.* In letters where there is no interpersonal action, this "Sincerely yours" element can also be omitted. If the complimentary close contains more than one word, capitalize only the first word and punctuate with a comma.

- *Signature block.* This is the blank area for the signature, followed by your typed name, title, and organization. In professional correspondence, don't forget to include those letters that identify the degree or title that you worked so hard to earn. Below your name, include your title and the name of your company or organization.

- *End notations.* These elements are the "Cc:" and "Encl:" abbreviations below the signature block. The first set is the initials of the sender and typist, respectively (for example, "JMC/rbs"). Labels such as "Encl.," "Enclosure," or "Attachments" indicate that other documents have been attached to the letter. If you want, you can specify exactly what you've attached: for example, "Encl.: specifications."

  "Cc:" followed by one or more names indicates to whom a copy of the letter is sent. "Bcc:" is an office-politics stratagem that identifies "blind" recipients. If you receive a letter with "Bcc:" at the bottom, the people whose names follow "Bcc:" do not know that you received the letter, nor do they know that you know that they received the letter.

- *Following pages.* If you use letterhead stationery, use the matching stationery (the same quality and style of paper but without the letterhead) on following pages. On following pages in professional correspondence, use a header like

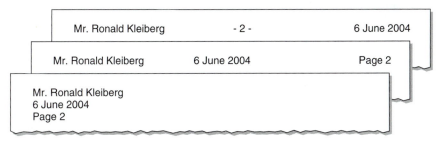

Figure 4-2 Three separate formats for following pages in business letters.

one of those shown in Figure 4-2, in which you include the name of the addressee, the date, and the page number.

COMMON BUSINESS-LETTER FORMATS

Traditionally, business letters have used one of four standard formats: block, semi-block, alternative block, and, more recently, simplified formats. These formats vary

Clarkson Hall, Rm. 1709
Monash University
Clarkson, WA 98881

25 May 2004

Hughes, Gano, Associates
1118 The High Road
Austin, TX 78703

Dear Colleague:

I am writing to professional consultants like yourself in an attempt to survey any experience you may have using different dynamic solvers to solve undamped linear eigenproblems, particularly large eigenproblems (with greater than 5000 degrees of freedom).

A wide variety of solvers is available. They vary from subspace iteration, to Lanczos, to conjugate-gradient, to dynamic condensation, and to component mode synthesis. I would like to know what professional engineers are using and why. (For example, is your choice faster or more robust, or do you have some other criteria?)

At present, my interests are focused on solving unsymmetric linear equations in the boundary element method. However, from a practical viewpoint, I have attempted to solve a liquid oscillation problem by using "pseudo-fluid" elements in NISA. The trouble involves choosing a Poisson's ratio as close to 0.5 as possible.

I found the Lanczos method to be best in this case, but there were other difficulties in simulating the boundary conditions at the top of the fluid with appropriate springs. For these reasons, I ultimately had to abandon this idea. Even so, the subspace and accelerated subspace iteration techniques were not nearly as effective.

Currently, I am doing research in the area of accelerating the solution of linear undamped eigenproblems and am interested in comparing what actual users find most useful (and not just the theoretical researchers!).

I would very much appreciate hearing about any experience or insights you may have had in these areas. If it would be easier for you, you can contact me by e-mail; my address is janemc@pink.cc.monash.edu.us.

Sincerely,

Jane A. McMurrey

Jane A. McMurrey, P.E., Ph.D.

Figure 4-3   Standard business-letter format—semiblock letter.

according to which elements are present (for example, a salutation) and where they are placed on the page (for example, on the left or right margin).

Figures 4-1, 4-3, and 4-4 show these business-letter formats:

- *Block format*: The easiest and most commonly used; all elements are flush left.
- *Semiblock format*: Similar to the block format except that the heading, complimentary close, and signature block are on the right margin.
- *Alternative block format*: The same as the semiblock format except that it adds a subject line.
- *Simplified format*: The same as the block format except that it omits the salutation.

For communications that involve no interpersonal interaction, the simplified and the alternative formats are acceptable. Notice that the letter for the job announcement in Figure 4-4 uses the alternative format. (It could also have used the simplified format.) However, for serious professional communications, such as proposals or employment letters, stick with the block format.

---

25 May 2004

Dr. Patrick H. McMurrey
Department of Mechanical Engineering
Clarkson Hall, Rm. 1709
Monash University
Clarkson, WA 98881

SUBJ.: Position for experienced development engineer

Dear Colleague:

CSMI is seeking qualified development engineers. Please distribute this letter to anyone in your organization who might be interested in working with us.

CSMI is a leading sawmill equipment manufacturer headquartered in Portland, with manufacturing facilities in Portland and Hot Springs, AR. We are looking for a seasoned (8 to 10 years) development engineer with a hands-on style and a strong background of stress analysis and design optimization for large capital equipment. A bachelor's degree in mechanical engineering is required; an advanced degree is preferred.

CSMI offers competitive compensation, company-paid health, dental, life and pension Optional 401(k). CSMI is a drug-free workplace. We are also an equal-opportunity employer; qualified applicants who would enhance our cultural diversity are encouraged to apply.

To be considered, please submit a resume with salary history and requirements to:

Human Resources Manager
CSMI
4000 NW St. Helens Rd.
Portland, OR 97210

Figure 4-4   Standard business-letter format—alternative block letter. This format includes a subject line and omits the complimentary close.

REPORT-LIKE LETTERS AND COVER LETTERS

If you are writing a report or some other "standalone" document, attach a cover letter to the front when you send it. This is a brief business letter that announces what the report is about, why it was written, for whom, and other such identifying details.

However, if the report is short—two to three pages, for example—you can incorporate it right into the framework of a business letter (or memo, if it's internal). You can present an engineering report—complete with tables, illustration, lists, and headings—within the confines of a business letter. For an example, see Figure 5-5.

BUSINESS MEMORANDA

For communications internal to an organization, use the memorandum format. Examples of such communications are a call for employees to attend a general meeting, a reminder that status reports are due, the actual text of a status report, a request to an employee to provide information, and that employee's subsequent report of the information. The actual contents of a memo can be very much like those of a business letter or like those of a short report—the key is the memorandum format.

STANDARD COMPONENTS OF MEMORANDA

Memorandum format is much simpler than that of business letters. Figures 4-5 and 4-6 illustrate the standard components.

- *Heading—DATE.* Although formats vary, put the date you send the memo in the header. The example in Figure 4-6 shows it as the third line in the header; in some designs it is the first line, as in Figure 4-5.
- *Heading—TO.* Put the name of the recipient or the group name in this slot. The level of formality is very apparent here. You can put "Sarah," "Sarah James," or "Ms. Sarah James, Director of Personnel," depending on your familiarity with the recipient and the formality of the situation.
- *Heading—FROM.* Put your own name or the name of the person or group for whom you are writing the memo in this slot. Once again, familiarity and formality dictate whether to put just your first name, your full name, or your full name and title. As the writer of the memo, jot your initials or first name just after your printed name.
- *Heading—SUBJECT.* In this slot, place a phrase that captures the topic and purpose of the memo. For a survey of grammar-checking software, the subject might be "Results of our survey on grammar-checking software." The actual label for this element varies: Some styles use "RE:" or "SUBJ.:". If your memo is in response to something, phrase the subject line accordingly.

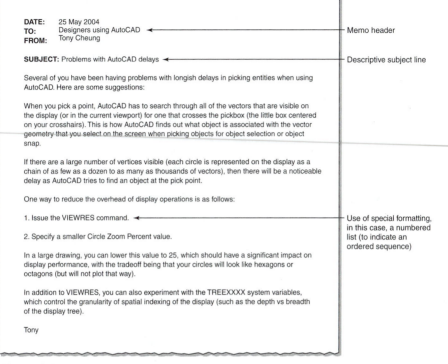

DATE: 25 May 2004
TO: Designers using AutoCAD ◄──────────────────────── Memo header
FROM: Tony Cheung

SUBJECT: Problems with AutoCAD delays ◄──────────────────── Descriptive subject line

Several of you have been having problems with longish delays in picking entities when using
AutoCAD. Here are some suggestions:

When you pick a point, AutoCAD has to search through all of the vectors that are visible on
the display (or in the current viewport) for one that crosses the pickbox (the little box centered
on your crosshairs). This is how AutoCAD finds out what object is associated with the vector
geometry that you select on the screen when picking objects for object selection or object
snap.

If there are a large number of vertices visible (each circle is represented on the display as a
chain of as few as a dozen to as many as thousands of vectors), then there will be a noticeable
delay as AutoCAD tries to find an object at the pick point.

One way to reduce the overhead of display operations is as follows:

1. Issue the VIEWRES command. ◄─────────────────────────── Use of special formatting,
                                                            in this case, a numbered
2. Specify a smaller Circle Zoom Percent value.   list (to indicate an
                                                            ordered sequence)
In a large drawing, you can lower this value to 25, which should have a significant impact on
display performance, with the tradeoff being that your circles will look like hexagons or
octagons (but will not plot that way).

In addition to VIEWRES, you can also experiment with the TREEXXXX system variables,
which control the granularity of spatial indexing of the display (such as the depth vs breadth
of the display tree).

Tony

Figure 4-5   Example of a business memorandum.

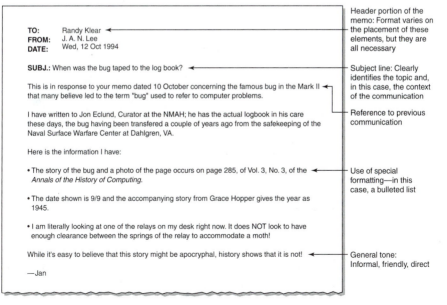

                                                            Header portion of the
                                                            memo: Format varies on
TO: Randy Klear ◄──────────────────────────────────── the placement of these
FROM: J. A. N. Lee   elements, but they are
DATE: Wed, 12 Oct 1994   all necessary

SUBJ.: When was the bug taped to the log book? ◄─────────── Subject line: Clearly
                                                            identifies the topic and,
This is in response to your memo dated 10 October concerning the famous bug in the Mark II ◄ in this case, the context
that many believe led to the term "bug" used to refer to computer problems.   of the communication

I have written to Jon Eclund, Curator at the NMAH; he has the actual logbook in his care ──── Reference to previous
these days, the bug having been transfered a couple of years ago from the safekeeping of the communication
Naval Surface Warfare Center at Dahlgren, VA.

Here is the information I have:

• The story of the bug and a photo of the page occurs on page 285, of Vol. 3, No. 3, of the ◄─ Use of special
  Annals of the History of Computing.   formatting—in this
                                                                                              case, a bulleted list
• The date shown is 9/9 and the accompanying story from Grace Hopper gives the year as
  1945.

• I am literally looking at one of the relays on my desk right now. It does NOT look to have
  enough clearance between the springs of the relay to accommodate a moth!

While it's easy to believe that this story might be apocryphal, history shows that it is not! ◄─ General tone:
                                                                                               Informal, friendly, direct
—Jan

Figure 4-6   Example of a business memorandum.

For example: "Re: your request for a grammar-checker survey" or "Review of your grammar-checker survey results."

- *Signature block.* In formal memoranda styles, writers actually insert the same kind of complimentary close and signature block that you see in business letters.

REPORT-LIKE MEMOS AND COVER MEMOS

When you write a short report (for example, under three pages), you can put the entire report into the memo—headings, lists, graphics, the works. Or you can create a cover memo and attach the report as a separate document. The cover memo briefly announces the topic and purpose of the attached information, provides an overview of its contents, and requests something, for example, a review or response. (See Figures 5-1, 5-5, 5-6, and 5-7.)

EMAIL

These days, email may seem like the only communication tool you need for professional work. In fact, email has become such a crucial job skill that it is required by many employers, even though you don't see it listed in job descriptions. The following focuses on email functions, style, and format.

IMPORTANT EMAIL FUNCTIONS

In the first edition of this book, email was a new enough phenomenon that it was appropriate to list the fundamental email skills people should master. But since then, email has become as commonplace as the telephone. Still, if you use email in professional work, make sure you are comfortable with the following not-so-basic skills:

- *Save email into files or folders.* Organize your sent and received email into meaningful folders—for example, "clients," "staff," "projects," "friends & family."
- *Keep copies of email you send.* Don't delete the email you send. Sometimes it disappears into the Internet void, recipients may accidentally lose it, or you can't remember what you wrote.
- *Search email folders.* Know how to search your email folders for topics or for the names of recipients or senders. Inevitably, you'll forget what you wrote to a client or what that client wrote to you. Doing a quick search is far better than scanning through hundreds of emails.

- *Create and use aliases and distribution lists.* Increase your email efficiency by creating aliases (short abbreviations for email addresses) and distribution lists (groups of related email addresses, such as those for staff or customers).

- *Use a signature.* If you need to include your full name, title, organization, phone and fax numbers, and other such information in your email, set up a "signature"—it automatically pops that information into every email you send.

- *Use templates.* If some of your email has standard content, set up templates. For example, create a template for your standard request for bids or announcement of services, import the template into your email, and just change the necessary details.

- *Attach files to email.* Learn how to attach electronic files to email—for example, reports written in Microsoft Word or spreadsheets created in Lotus 123.

- *Proofread and spellcheck your email.* Learn how to spellcheck your email, and get into the habit of spellchecking every email you send. Proofread your email, checking specifically for missing words—which a spell checker cannot catch. (Imagine the devastating effects of leaving out a single "not.")

- *Plan how to access your email on the road.* Know how to access email while on business trips: Use an Internet service such as Earthlink that has local numbers throughout the United States, use a free email service such as Yahoo or Hotmail, or use software such as Symantec's pcAnywhere or Expertcity's GoToMyPC that enables you to log into your office or home computer and work as if you were sitting right there.

EMAIL FORMAT AND STYLE

In the early days of email, all you had was plain text. Now you can include different fonts, different type sizes, different color, graphics, tables, and even animation in your email. In other words, anything you can do in a print report, letter, or memo, you can do in email!

As for style in email messages, here are some suggestions:

- *Informality.* Adjust the tone of your email according to the recipients and situation. Informality is common in email, but think twice about using humor, sarcasm, and informality with business clients and higher-level management—especially those whose native language is not English.

- *Brevity.* Email messages are normally rather short—for example, under a dozen lines—and the paragraphs are short as well. No one likes having to do a lot of extended reading on a computer screen. Consider putting lengthy messages in printable documents and attaching them to your email. True, you can print email, but the pagination is faulty.

- *Specific subject lines.* To ensure email gets read and has the desired impact, make the subject line specific and compelling. If recipients have 60 to 70 messages waiting in the inbox and all they can see is the subject lines, they are more likely to read email that indicates a specific business topic.

- *Important information first.* Heavy email users tend to lose interest or patience quickly. Put the most important information at the beginning of your message. Use subsequent sentences for elaboration.

- *Short paragraphs and space between paragraphs.* Whenever possible, break your messages into paragraphs of less than six or seven lines.[1] When you divide your message into paragraphs, skip an extra line between them.

- *Highlighting and emphasis.* Contemporary email now enables you to use typographical effects (bold, italics, color, different fonts), tables, and graphics. Your job is to use typography consistently and in moderation (see Chapter 2). You can also use tables and graphics to reinforce your messages (see Chapter 7).

- *Headings.* If your email requires readers to press the PageDown key more than twice, use headings to identify the subtopics within the message. For these headings, use a slightly larger font and bold type. (See Chapter 2 for details on headings.)

- *Lists.* If you have key points to emphasize, if you make a series of points, or if you are presenting step-by-step information, use the various forms of lists that are available. As shown in Figure 4-7, contemporary email software now provides automatic formatting for different types of lists. (See Chapter 2 for more on lists.)

- *Automatic replies.* The reply function in email is a great time saver. However, email is often addressed to multiple recipients. Imagine that in replying to a partner on a project, you question the competence of another partner in that same project. But what if the first individual had copied (Cc:) that partner on his original note, and your reply also goes to that other, first individual? Uh oh . . .

WRITING STYLES FOR BUSINESS CORRESPONDENCE

Regardless of the medium you use for your business correspondence, most of the guidelines for writing style are the same. Whether you are writing a business letter, memorandum, or email, the following recommendations are equally valid.

---

[1] In an informal sampling, David Crystal found that 80% to 90% of email paragraphs were five lines or less (*Language and the Internet.* Cambridge: Cambridge University Press, 2001, p. 115).

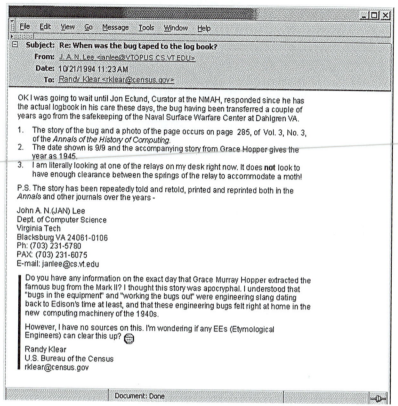

Figure 4-7   Email example. In this exchange, Mr. Klear is looking for information on the famous bug that got into an early computer and thus gave rise to the computer term. In his response, Dr. Lee appends the text of Mr. Klear's original message.

- *Indicate the topic in the first sentence.* Don't assume recipients read your subject line (however clear and compelling it may be). State the topic and purpose of your communication in the very first sentence.
- *Identify any situation or preceding correspondence to which your communication responds.* In the first paragraph, establish the context by referring to any previous meeting, phone conversation, or correspondence.
- *Provide an overview of the contents of the communication.* If the letter, memo, or email is lengthy, provide an overview of the contents—nothing more than an informal list in a sentence within the first paragraph.
- *Keep the paragraphs short.* Ideally, paragraphs in business correspondence should not go over six to eight lines. In business correspondence, readers are less willing to wade through long, dense paragraphs than they are, for example, in textbooks or formal reports. (However, don't start new paragraphs just anywhere. Divide your communication into paragraphs at those points where the topic changes.)

- *Use headings for communications over a page in length.* If your communication is more than a page or two and if the information in it is like that in a report, use headings to mark off the boundaries where new topics start. (See Chapter 2 for more on headings.)

- *Use lists and graphics as you would in a report.* Business correspondence can at times resemble reports: Writers use the same sorts of headings, lists, and graphics in their letters and memos. Look for ways to create lists, particularly in long paragraphs. Similarly, use graphics and tables in your correspondence just as in regular reports.

- *Be brief, succinct, to the point.* Brevity is never so important as it is in business correspondence—and still more so in email. Readers lack patience with unnecessary background and wordiness.

- *Use an interactive style in memos and email.* Be as informal as the situation allows. Whenever appropriate, use the "you" style of writing—avoid the impersonal third-person and passive-voice styles.

- *Indicate any action necessary on the part of the recipient.* Let readers know what you expect them to do as a result of reading your correspondence. What actions should they take after reading your letter, memo, or email? Should they fill out a questionnaire? Where is it located? Where should they send it? Make sure that all details like these are clearly and specifically explained.

EXERCISES

Talk to several professional engineers about the business correspondence they write or receive:

1. What are the typical audience, purpose, and content of their letters and memos? Why letters and memos as opposed to phone calls?

2. How much secretarial assistance do they receive? Do they get any help editing or proofreading their correspondence?

3. When they have to convey specialized, technical information, is it to another engineer, or often must they translate for nonspecialists?

4. How do they decide between writing a hardcopy letter or memo, making a phone call, or sending email?

5. What do they see as the advantages and the problems of using email in conducting their business?

BIBLIOGRAPHY

Angell, David, and Brent Heslop. *The Elements of E-Mail Style*. Reading, MA: Addison-Wesley, 1994.

Blake, Gary. *Quick Tips for Better Business Writing*. New York: McGraw-Hill, 1995.

Crystal, David. *Language and the Internet*. Cambridge: Cambridge University Press, 2001.

Extejt, M. M. "Teaching Students to Correspond Effectively Electronically." *Business Communications Quarterly* 61, no. 2 (June 1998), 56–67.

Gates, Bill. *Business @ the Speed of Thought*. New York: Warner, 1999.

Kanter, Rosabeth M. *Evolve! Succeeding in the Digital Culture of Tomorrow*. Boston, MA: Harvard Business School Press, 2001.

Krol, Ed. *The Whole Internet: User's Guide and Catalog*. Sebastopol, CA: O'Reilly, 1994.

Lamb, Linda, and Jerry D. Peek. *Using E-Mail Effectively*. Sebastopol, CA: O'Reilly, 1995.

Perry, T. S. "E-Mail at Work." *IEEE Spectrum* (October 1992), 24–28.

Perry, T. S., and J. A. Adam. "E-Mail Pervasive and Persuasive." *IEEE Spectrum* (October 1992), 22–23.

5

WRITING COMMON ENGINEERING DOCUMENTS

Documentation—preparing reports, manuals, procedures, proposals, papers, even marketing support materials—is an essential part of the engineering job. Increasingly, it's part of the engineer's job. What used to be done by secretaries, inhouse publication functions, or outsource contractors must now be done, or at least managed, by the performing engineer.

Companies, as well as public institutions, are trying to get "lean and mean." This means less secretarial and clerical help, fewer engineering technician and entry-level engineers, and little or no inhouse publication service to handle the details of document preparation and production.

Joan Nagel, *Handbook for Preparing Engineering
Documents* (New York: IEEE Press, 1996), p. 1

This chapter explores some of the common types of reports you may write as an engineer, particularly in terms of their typical content and organization. As you read this chapter, keep in mind that the names of these types vary considerably, and their contents often combine in different ways.

- *Inspection or trip reports:* Briefly report on the inspection of a site, facility, or property; summarize a business trip; or report on an accident, describing the problem, discussing the causes and effects, and explaining how it can be avoided.
- *Laboratory and field reports:* Report on an experiment, test, or survey; present the data collected and discuss the research theory, method, or procedure;

discuss conclusions, and, possibly, explore applications of the findings or possibilities for further research.

- *Specifications:* Provide detailed requirements for a product to be developed or detailed descriptions of an existing product; provide specifics on design, function, operation, and construction.

- *Proposals:* Seek a contract, approval, or funding to do a project; function as a competitive bid to get hired to do a project; promote you or your organization as a candidate for a project; promote the project itself, showing why it is needed.

- *Progress reports:* Summarize how your project is going, what you or your group has accomplished, what work lies ahead, what resources have been used, what problems have arisen.

- *Instructions:* Explain how to perform certain tasks, provide procedures on using equipment, give troubleshooting and maintenance guidelines, explain policies and operating procedures.

- *Recommendation reports:* Study a situation or problem, report on various alternatives or options, recommend the best one or assess the feasibility of an idea.

Note   The reports discussed in this chapter are mostly short and informal and are routinely formatted as in-office memoranda or business letters. However, practically any of these reports can be formatted as full-length formal reports. (See Chapter 6 for the design of full-length reports.)

SOME PRELIMINARIES

Before getting into the details, consider some points that apply to all the types of reports about to be discussed:

- *Don't get hung up on the names of reports.* Sorry, there is no ANSI standards committee on the proper names, contents, and format of reports. Don't worry about whether a report is really an evaluation report or really a recommendation report. Determine the requirements for the report you must write; think as clearly as you can about the needs and requirements of the audience of that report. One or some combination of the report types discussed here is likely to suffice.

- *Find out your company's requirements.* This chapter illustrates common formats, contents, and organization for reports. However, every company, organization, field, and profession has its own names for reports as well as its own requirements on format, content, and organization. Those requirements may be written down in some official document, or they may be traditions that everybody "just knows." Your job is to find out what those requirements or traditions are. The discussion and examples in this chapter give you some clues about what to expect, and something to use when there are no guidelines.

- *Think creatively about content and organization.* Rarely will the contents and organization of the reports described here be a perfect fit for your real-world report projects. The plans for reports presented here cannot be used as templates. Always think creatively, brainstorming about what else your readers may need and what else the report-writing situation calls for.

- *Build your reports on the needs of your audience.* Everything about your report depends on the specific people who are going to read it. Sometimes you must write for different audiences within the same report. See Chapter 2 for a detailed discussion of analyzing and adapting to audiences.

- *Be careful with discussion of technical background.* Most of the reports in this chapter require technical background—but not necessarily. Background sections provide information to make the rest of the report understandable. However, loosely related background sections are not helpful. To avoid this common problem, write the main text of your report first, and then review it for what readers may need help with; only afterward write the background section, with the readers' needs squarely in mind.

- *Be careful with the report introduction.* Another problem concerns introductions. An introduction introduces readers *to the report*—not the technical subject matter. The introduction gets the reader ready to read the report. It announces the topic, alludes to the situation that brought about the need for the report, indicates what the audience needs to know to understand the report, and provides a brief overview of the topics to be covered (and not covered). It's bad practice to dive right into the main subject matter in an introduction—readers then lack any perspective, overview, or roadmap for the whole report.

INSPECTION AND TRIP REPORTS

One common group of engineering reports handles tasks such as reporting on the inspection of a site or facility; describing an incident or accident and exploring the causes, effects, and prevention; summarizing the events and results of a business trip; and describing property, equipment, or new technology. You might hear these types of reports referred to variously as site reports, inspection reports, incident reports, trip reports, or accident reports. (See Chapter 1 for many more names for reports.)

These reports are similar in that they contain lots of description, narration, and discussion of related causes and effects, as well as a certain amount of interpretation and evaluation. If you report on an accident, you describe the damage, then explore the causes. If you report on a business trip, you narrate the events; if the purpose of the trip was to assess a new technology, you also do some evaluation. Obviously, these types of short reports overlap considerably. It's a loose category—the names are by no means widely agreed upon:

- *Trip reports:* Discuss the events, findings, and other aspects of a business trip. This type documents observations so that people in your organization can share them (see Figure 5-1).
- *Investigation or accident reports:* Describe your findings concerning a problem; explore its causes, its consequences, and what can be done to avoid it.
- *Inspection or site reports:* Report your observations of a facility, a property, or an installation of equipment, with description and possibly evaluation of it.

Many reports like these are now composed and transmitted strictly as email. For example, if you had been to a conference on new surface mount techniques, you'd want to make these discoveries available to colleagues back at work. Upon returning from the conference, you could send everyone email or, better still, put your report in some electronically accessible location.

Contents and Organization

For the content of your informal report, consider these suggestions.

Introduction. No matter which type of report you write, begin by indicating the purpose of the report and providing a brief overview of its contents. Avoid that impulse to dive right into the thick of the discussion!

Background. It's also a good idea to set the stage—to explain the background or context of the report. Why did you go on this business trip? Why were you sent to inspect the facility? Who sent you? What are the basic facts of the situation—the time, date, place, and so on?

Factual Discussion. The main contents of a report like this are some combination of description or narration. Typically, you must describe the accident, facility, property, or the proposed equipment. Give an account of what happened on the trip: where you went, who you met with, and what was discussed.

Actions Taken. If you are investigating a problem and implementing a solution, your report should contain a step-by-step discussion of how you determined the problem and corrected it.

Interpretive, Evaluative, or Advisory Discussion. Once you've laid the foundation with the background and factual discussion, you're set to do what readers may expect—to evaluate the property or equipment, explain what caused the accident, interpret the findings, suggest further action, or recommend ways to prevent the problem in the future.

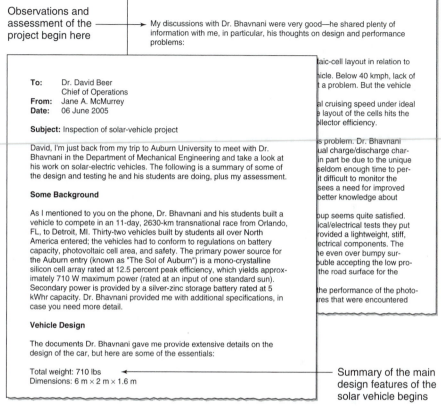

Observations and assessment of the project begin here ──►

My discussions with Dr. Bhavnani were very good—he shared plenty of information with me, in particular, his thoughts on design and performance problems:

To: Dr. David Beer
Chief of Operations
From: Jane A. McMurrey
Date: 06 June 2005

Subject: Inspection of solar-vehicle project

David, I'm just back from my trip to Auburn University to meet with Dr. Bhavnani in the Department of Mechanical Engineering and take a look at his work on solar-electric vehicles. The following is a summary of some of the design and testing he and his students are doing, plus my assessment.

Some Background

As I mentioned to you on the phone, Dr. Bhavnani and his students built a vehicle to compete in an 11-day, 2630-km transnational race from Orlando, FL, to Detroit, MI. Thirty-two vehicles built by students all over North America entered; the vehicles had to conform to regulations on battery capacity, photovoltaic cell area, and safety. The primary power source for the Auburn entry (known as "The Sol of Auburn") is a mono-crystalline silicon cell array rated at 12.5 percent peak efficiency, which yields approximately 710 W maximum power (rated at an input of one standard sun). Secondary power is provided by a silver-zinc storage battery rated at 5 kWhr capacity. Dr. Bhavnani provided me with additional specifications, in case you need more detail.

Vehicle Design

The documents Dr. Bhavnani gave me provide extensive details on the design of the car, but here are some of the essentials:

Total weight: 710 lbs ◄────
Dimensions: 6 m × 2 m × 1.6 m

taic-cell layout in relation to
hicle. Below 40 kmph, lack of
t a problem. But the vehicle

al cruising speed under ideal
e layout of the cells hits the
ollector efficiency.

s problem. Dr. Bhavnani
ual charge/discharge char-
in part be due to the unique
seldom enough time to per-
it difficult to monitor the
sees a need for improved
better knowledge about

oup seems quite satisfied.
ical/electrical tests they put
rovided a lightweight, stiff,
ectrical components. The
he even over bumpy sur-
ouble accepting the low pro-
the road surface for the

the performance of the photo-
ures that were encountered

──── Summary of the main design features of the solar vehicle begins

Figure 5-1 Short business-trip report—excerpts. The writer summarizes her visit with researchers involved in the solar-vehicle design and provides an assessment of that work. (Information for this report was developed from S. H. Bhavnani, "Design and Construction of a Solar-Electric Vehicle," *Journal of Solar Energy Engineering* (February 1994), pp. 28–34).

FORMAT

Just because it is a short and informal report, don't neglect to use basic formatting practices that will make your report more readable, more usable, and more accessible—not o mention more professional in appearance.

- Unless the report goes over several pages or unless your company has certain requirements, use the memorandum format. (And obviously, if you are writing to an individual or organization external to your company, use the business-letter format.)
- Use headings to mark off the major subtopics within the report. Notice how they are used in the example report in Figure 5-1. Headings help readers skip to the sections they want to read.

- Use the various types of lists as needed. These help emphasize key points, make information easier to follow, help readers return to key points, and generally create more white space—all of which makes your report more readable.
- Use tables and graphics as necessary.

LABORATORY AND FIELD REPORTS

Laboratory and field reports present not only the data from an experiment or survey and the conclusions that can be drawn from that data, but also the theory, methods, procedures, and equipment used in that experiment or survey. (See the excerpts from a laboratory report in Figure 5-2.)

CONTENTS AND ORGANIZATION

As much as practical, the laboratory or field report should enable readers to replicate the experiment so that they can verify the results for themselves. Because of this dual requirement, laboratory and field reports have a characteristic structure.

Introduction.   In the introduction, indicate the overall topic and purpose of the report, and provide an overview of its contents. Remember: Avoid diving into the thick of the discussion; orient readers to the report topic, purpose, and contents first.

Background.   Provide a discussion of the background leading up to the project. Typically, this involves discussing a research question or conflicting theories in the research literature. Or, for example, you may want to apply an interesting discovery from another field to something in your own. Explore this background to enable readers to understand why you are doing this work. When you do, provide citations for the sources of information you use, using the standard bibliographic format (see Chapter 11).

Literature Review.   Often included in the lab or field report is a discussion of the research literature related to your project. You summarize the findings of other researchers that have a bearing on your work. Again, use the standard bibliographic format.

Depending on the length and complexity of the report, all three of the elements just discussed—introduction, background, and literature review—may easily combine into one introductory paragraph without subheadings. Regardless of their length, these three elements should occur at the beginning of a laboratory or field report, even if each is only a sentence or two.

The data—the findings—from the research are presented. Tables, charts, and graphs can be used to show the relationships and trends more vividly. (Large tables can be shifted to an appendix.)

Background on the project: The problem is introduced, and related research is summarized.

Introduction

The increasing use of plastic films for co[...] drink calls for more information concernin[...] plastic packaging materials with food an[...]

During droughts, it is a common problem to make local water potable and to store taste and odors are known to develop in after direct exposure to sunlight for long for these organoleptic changes have bee[...]

In fact, it is often the transfer of materials aging that is the origin of off-flavors in fin[...] more, plastic packaging film is often prim[...] ual solvents such as hydrocarbons, alcol hens, et al. 1984) into the plastic. These packaged food (Kim and Gilbert, 1989) a because of their low flavor thresholds (H[...]

This study reports on the concentrations pounds released into drinking water sam[...] and printing ink.

Experimental Section

Local well water was used for this work unless otherwise stated. Polyethylene (PE) was an Enichem product. HPLC-grade water was a Merck product. Horseradish peroxidase was a Sigma product. Samples were stored in a well-aerated dark room and were analyzed after 15 days. The exposition to direct sunlight occurred when the samples were put on the roof of the building for 15 days in June.

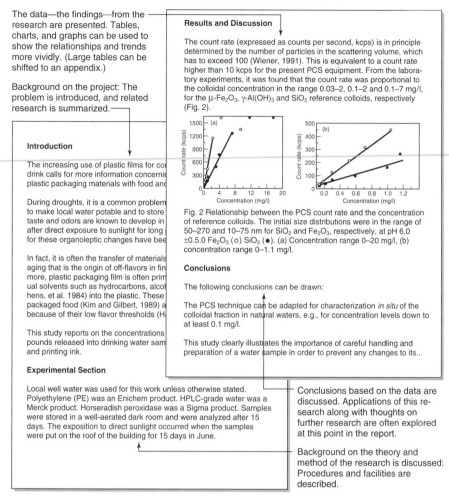

Results and Discussion

The count rate (expressed as counts per second, kcps) is in principle determined by the number of particles in the scattering volume, which has to exceed 100 (Wiener, 1991). This is equivalent to a count rate higher than 10 kcps for the present PCS equipment. From the laboratory experiments, it was found that the count rate was proportional to the colloidal concentration in the range 0.03–2, 0.1–2 and 0.1–7 mg/l, for the μ-Fe_2O_3, γ-$Al(OH)_3$ and SiO_2 reference colloids, respectively (Fig. 2).

Fig. 2 Relationship between the PCS count rate and the concentration of reference colloids. The initial size distributions were in the range of 50–270 and 10–75 nm for SiO_2 and Fe_2O_3, respectively, at pH 6.0 ±0.5.0 Fe_2O_3 (o) SiO_2 (●). (a) Concentration range 0–20 mg/l, (b) concentration range 0–1.1 mg/l.

Conclusions

The following conclusions can be drawn:

The PCS technique can be adapted for characterization *in situ* of the colloidal fraction in natural waters, e.g., for concentration levels down to at least 0.1 mg/l.

This study clearly illustrates the importance of careful handling and preparation of a water sample in order to prevent any changes to its...

Conclusions based on the data are discussed. Applications of this research along with thoughts on further research are often explored at this point in the report.

Background on the theory and method of the research is discussed: Procedures and facilities are described.

Figure 5-2 Laboratory report excerpts with background, research methods, data, and conclusions. (Excerpts on the plastic-packaging experiment were drawn from Lucia Calvosa et al., "Taste and Odor Development in Water in Polyethylene Containers Exposed to Direct Sunlight," *Water Research* (July 1994), pp. 1595–1600. Excerpts from the study of colloidal matter in groundwater were drawn from Anna Ledin et al., "Measurements *in Situ* of Concentration and Size Distribution of Colloidal Matter in Deep Groundwaters by Photon Correlation Spectroscopy," *Water Research* (July 1994), pp. 1539–1545.)

Theory, Methods, Procedures, and Equipment. The next major section in the laboratory or field report presents your theory or approach in relation to your project. For example, as a software engineer, you may suspect that computer users would prefer online documentation to printed documentation. To test this idea, you set up several computers in a laboratory and have a typical cross section of computer users perform procedures you design. First, you'd discuss the common thinking on this subject—that computer users prefer printed material. Then you'd explain your method and procedures as well as the equipment and facilities. Readers could use this part of the report in particular to replicate your project.

Observations, Data, Findings, or Results.   In this type of report, you collect data and then organize and present it in a section of its own. The common approach is to present the data, often formatted into tables, graphs, or charts, without interpretive discussion.

Conclusions.   In the conclusions section of a laboratory or field report, you draw conclusions based on the data you've gathered and explain why you think those conclusions are valid.

Implications and Further Research.   Laboratory and field reports also typically explore the implications of conclusions, considering how they can be applied and outlining further research possibilities.

As with the opening sections, these three sections—findings, conclusions, and implications—can be rolled into one if the report is brief and relatively simple. In whatever way they are combined, the first two elements—the data and the conclusions—must be there.

Information Sources.   Most laboratory and field reports conclude with a section that lists information sources used in the project. For entries in that list, use the bibliographic format shown in Chapter 11.

FORMAT

The laboratory or field report can be presented in memorandum format if it is short and addressed internally within an organization, or it can be presented as a formal report, with covers, table of contents, and appendixes. For reports over three or four pages, consider using the formal-report format (see Chapter 6).

SPECIFICATIONS

Specifications are descriptions of products or product requirements. They provide details for the design, manufacture, testing, installation, and use of a product. You typically see specifications in the documentation that comes with certain kinds of products, for example, DVD players or computers. These describe the key technical characteristics of those items. But specifications are also written as a way of "specifying" construction and operational details. They are then used by people who actually do the construction or purchasing. When you write specifications, accuracy, precision of detail, and clarity are critical. Poorly written specifications can cause a range of problems and lead to lawsuits.

For these reasons, then, specifications have a particular style, format, and organization. If you write specifications, find out the specific requirements for format, style, contents, and organization for them. If these are not documented, collect a big pile of specifications written for your company, and study them. Some general recommendations follow.

- Use two-column lists or tables (as shown in Figure 5-3) to list specific details. If the purpose is to indicate details such as dimensions, materials, weight, tolerances, and frequencies, regular paragraph-style writing is not as effective.

- In sentence-style specifications, make sure each specific requirement has its own separate sentence and uses the decimal numbering system for ease of cross-referencing.

- In the sentence-style format, use an outline style similar to the one shown in Figure 5-3. Make sure that each specification receives its own number-letter designation so that each can be referenced separately.

- Use either the open (performance) style or the closed (restrictive) style, depending on the requirements of the job. In the *open* or *performance* style, you specify what the product or component should do, that is, its performance capabilities. In the *closed* or *restrictive* style, you specify exactly what it should be or consist of.

- Whenever possible, cross-reference existing specifications rather than repeating those details. Government agencies as well as trade and professional associations publish specifications standards. You can refer to these standards rather than copying them verbatim into your own specifications.

- Use specific, concrete language that identifies as precisely as possible what the product or component should be or do. Avoid ambiguity (using words that can be interpreted in more than one way). Use technical jargon the way it is used in the trade or profession.

- For specifications to be used in design, manufacturing, construction, or procurement, use *shall* to indicate requirements. In specifications writing, *shall* is understood as indicating a requirement. (See the outline-style specifications in Figure 5-3 for examples of this style of writing.)

- Provide numerical specifications in both words and symbols: for example, "the distance between the two components shall be three centimeters (3 cm)."

- Use a relatively terse writing style in specifications: Incomplete sentences are acceptable, as well as the omission of obvious function words such as articles and conjunctions.

- Exercise caution with pronouns and relational or qualifying phrases. There should be no doubt about the reference of words such as *it, they, which,* and *that.* Watch out for sentences containing a list of two or more items followed by

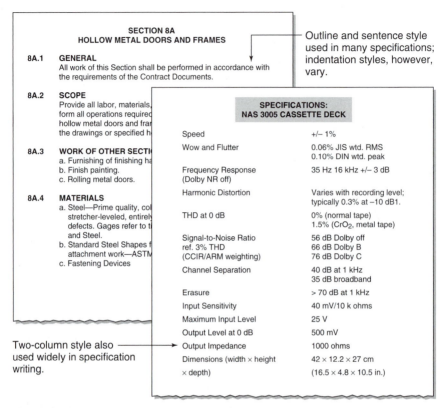

Figure 5-3 Specifications—excerpts. Outline and two-column style are commonly used to present information in specifications. Graphics, tables, and lists are heavily used, but some details can only be provided through sentences and paragraphs.

some descriptive phrase—does the descriptive phrase refer to all the list items or just one? In cases like these, use a wordier approach for the sake of clarity.

- Use words and phrases that have become standard in similar specifications over the years. Past usage has proven them reliable. Avoid words and phrases that are known not to hold up in lawsuits.

- Make sure your specifications are complete—put yourself in the place of those who need your specifications; make sure you cover everything they will need.

Test your specifications by putting yourself in the role of a bumbling contractor—or even an unscrupulous one. What are the ways a careless or incompetent individual could misread your specifications? Could someone willfully misread your specifications in order to cut cost or time? Obviously, no set of specifications can ultimately be foolproof or "shark-proof," but you must try to make them as clear and unambiguous as possible.

CONTENTS AND ORGANIZATION

Organization is critical in specifications—readers need to be able to find one or a collection of specific details quickly. To make individual specifications easy to find, use headings, lists, tables, and identifying numbers as previously discussed. Use one of the following organizational methods to facilitate quick retrieval.

- *General description.* Describe the product, component, or program first in general terms—administrative details about its cost, start and completion dates, overall description of the project, scope of the specifications (what you are not covering), anything general in nature that does not fit in the part-by-part descriptions.
- *Part-by-part description.* In the main text, present specifications part by part, element by element, trade by trade—whatever is the logical, natural, or conventional way of doing it.
- *General-to-specific order.* Wherever applicable, arrange specifications from general to specific.

GRAPHICS IN SPECIFICATIONS

In specifications, use graphics wherever they enable you to convey information more effectively. For example, in cleanroom specifications, drawings, diagrams, and schematics convey some of the information much more succinctly and effectively than sentences and paragraphs. See the example of a graphic used in specifications writing in Figure 5-4.

PROPOSALS

One of an engineer's most important tools, particularly a consulting engineer, is the proposal. With it, you get work, either for the company that employs you or for yourself, if you're an independent consultant.

If you explore the literature on proposals, you'll see them defined in different ways. In this book, however, the proposal is something quite specific: It is a bid, offer, or request to do a project plus any supporting information necessary to gain approval or acceptance to do the project. Proposals sometimes must convince the recipient that the project needs to be done, but proposals must always convince the recipient that the proposer is the right individual or organization to do the project.

In the typical proposal scenario, an organization[1] sends out a request for proposals (RFP) to do a project. These RFPs might be sent out for publication in newspapers,

---

[1] *Organization*, as the term is used here, refers to for-profit companies, not-for-profit organizations, and government agencies. All of these organizations, both commercial and noncommercial, request and submit proposals.

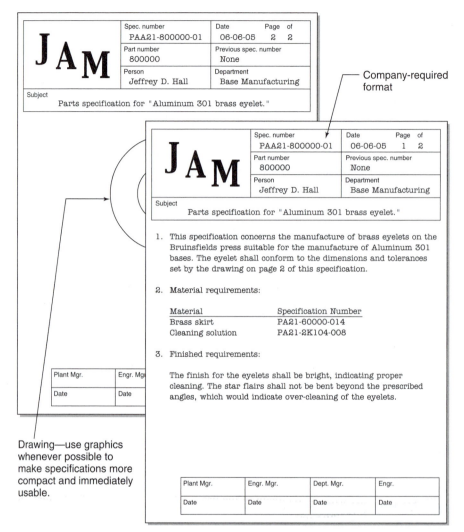

Company-required format

Drawing—use graphics whenever possible to make specifications more compact and immediately usable.

Figure 5-4 Graphics and tables used to present information in specifications. (Example specifications are drawn from work by Jeffrey D. Hall, engineering student, University of Texas at Austin.)

professional journals, or specialized periodicals such as the *Commerce Business Daily*; by mail to a select list of vendor organizations; or by various informal means such as telephone or email. Organizations then submit proposals in which they present their qualifications and make a case for themselves as a good choice. The recipient of the proposals selects one of the proposals and enters into contract negotiations. Once that is accomplished, the organization that wins the project can get down to work.

Proposal writing is a competitive affair. You must highlight your organization's strengths; you must make a good case for your company as the right one for the project.

TYPES OF PROPOSALS

Proposals are commonly divided into two types, based on whether the recipient requested them:

- *Solicited.* If an organization issues a request for proposals, the proposals are said to be *solicited*—they have been requested.
- *Unsolicited.* Individuals and companies often initiate proposals without formal requests from the recipients. They may see that an individual or organization has a problem or opportunity. When the proposal is unsolicited, you, the proposal writer, have to do the additional work of convincing the recipient that the project needs to be done.

Proposals can also be divided according to the context in which they occur:

- *Internal.* If you address your proposal to someone within your organization, the format and contents change significantly. The memo format is usually appropriate, and sections such as qualifications and costs may not be necessary.
- *External.* If you address your proposal to some other individual or organization outside your own, you must present your qualifications and use some combination of the business-letter and formal-report formats.

CONTENTS AND ORGANIZATION

The typical sections in a proposal are as follows (see the proposal excerpts in Figure 5-5 in which some of these sections are illustrated).

Introduction. In the first paragraph or section of a proposal, make reference to some prior contact with the recipient of the proposal or your source of information about the project. Identify the information that follows as a proposal (in other words, state the purpose). Also, give a brief overview of the contents of the proposal.

Background. In an unsolicited proposal, you should discuss the problem or opportunity that caused you to write the proposal. In solicited proposals, this may not be necessary: The party requesting proposals knows the problem very well. Still, a background section can be useful even in a solicited proposal: It demonstrates that you fully understand the situation and enables recipients to check your interpretation of it.

Actual Proposal Statement. Include a short section in which you state explicitly what you are proposing to do. Proposals often refer to many possibilities, which can create some vagueness about what's actually being offered. You may also need a scope statement—an explicit statement about what you are *not* offering to do.

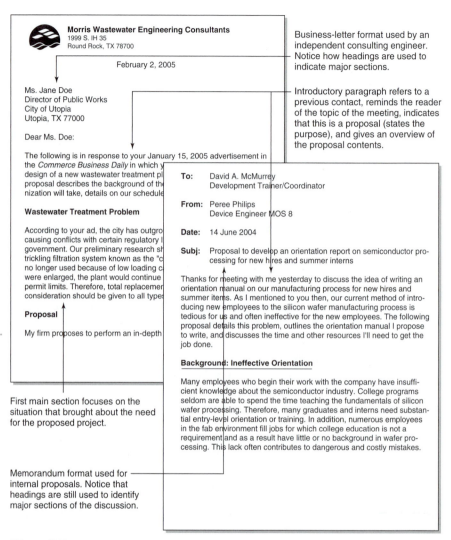

The following texts are embedded within the figure:

Morris Wastewater Engineering Consultants
1999 S. IH 35
Round Rock, TX 78700

February 2, 2005

Business-letter format used by an independent consulting engineer. Notice how headings are used to indicate major sections.

Ms. Jane Doe
Director of Public Works
City of Utopia
Utopia, TX 77000

Dear Ms. Doe:

Introductory paragraph refers to a previous contact, reminds the reader of the topic of the meeting, indicates that this is a proposal (states the purpose), and gives an overview of the proposal contents.

The following is in response to your January 15, 2005 advertisement in the *Commerce Business Daily* in which y[...] design of a new wastewater treatment pl[...] proposal describes the background of th[...] nization will take, details on our schedul[...]

Wastewater Treatment Problem

According to your ad, the city has outgro[...] causing conflicts with certain regulatory l[...] government. Our preliminary research sh[...] trickling filtration system known as the "c[...] no longer used because of low loading c[...] were enlarged, the plant would continue[...] permit limits. Therefore, total replacemen[...] consideration should be given to all type[...]

Proposal

My firm proposes to perform an in-depth[...]

First main section focuses on the situation that brought about the need for the proposed project.

Memorandum format used for internal proposals. Notice that headings are still used to identify major sections of the discussion.

To: David A. McMurrey
          Development Trainer/Coordinator

From: Peree Philips
          Device Engineer MOS 8

Date: 14 June 2004

Subj: Proposal to develop an orientation report on semiconductor processing for new hires and summer interns

Thanks for meeting with me yesterday to discuss the idea of writing an orientation manual on our manufacturing process for new hires and summer interns. As I mentioned to you then, our current method of introducing new employees to the silicon wafer manufacturing process is tedious for us and often ineffective for the new employees. The following proposal details this problem, outlines the orientation manual I propose to write, and discusses the time and other resources I'll need to get the job done.

Background: Ineffective Orientation

Many employees who begin their work with the company have insufficient knowledge about the semiconductor industry. College programs seldom are able to spend the time teaching the fundamentals of silicon wafer processing. Therefore, many graduates and interns need substantial entry-level orientation or training. In addition, numerous employees in the fab environment fill jobs for which college education is not a requirement and as a result have little or no background in wafer processing. This lack often contributes to dangerous and costly mistakes.

Figure 5-5   Proposal excerpts—one external, the other internal. These examples integrate the cover letter (or memo) and the proposal proper into one continuous document. (Example proposals are drawn from work by Peree Phillips and Christine Morris, students at Austin Community College.)

Description of the Work Product.   Some proposals need a section in which the proposed project—in other words, the results of the work—is described. This might be a constructed building, a program design, blueprints or plans, or even a 40-page report. The point is to provide details on what the recipients will get.

Benefits and Feasibility of the Project.   To promote the project to the recipient, some proposals discuss the benefits of doing the project. Others discuss the likelihood of those benefits. This is particularly true in unsolicited proposals, where the recipient must be convinced that the project is necessary in the first place.

Method or Approach.   Some proposals need a section that explains how you plan to go about the project, even the theory relating to your approach. For some projects, people need to know how the work will be done and why it will be done that way. As in the background section, this discussion enables you to demonstrate your professional expertise.

Qualifications and References.   Most proposals list the proposing organization's key qualifications, along with references to past work. This section is like a minire-sume. Large proposals actually include full resumes of the individuals who will work on the project. In internal projects where people know each other, this section is unnecessary.

Schedule.   The proposal should contain a schedule of the projected work, with dates or a timeline for the major milestones. This information may fit nicely in the methods and procedure section, or it may work better in a section of its own. Again, this gives the recipient an idea of what lies ahead and a chance to ask for changes; it also enables you to show how systematic, organized, and professional you are.

Costs.   Some proposals have a costs section that details the various expenses involved in the project. Rather than toss out a lump sum, break it down into different kinds of labor, hourly rates for each, and other charges. If you are writing an internal proposal, you may need to list supplies needed, expenses for new equipment, your time (even though it is not charged), and so on.

Conclusion.   Normally, the final paragraphs of your proposal urge the recipients to consider your proposal, contact you with questions, and of course accept your bid or request. This is also a good spot to allude once more to the benefits of doing the project.

FORMAT

You can package a proposal several ways, depending on your relation to the recipient, the size and nature of the proposal, and how the proposal will be used by the recipients.

- *Memorandum format.* If your proposal is short (under three pages), and if it's addressed to someone within your company, use a simple memo format (see Figure 5-5). Include headings as you normally would for any other document.
- *Business-letter format.* If your proposal is short but is addressed to someone outside your organization, use a business letter (also illustrated in Figure 5-5). Again, include headings as you normally would.
- *Separate proposal with cover memo.* If your proposal is long (over four pages), if it's addressed to people within your own company, and if you

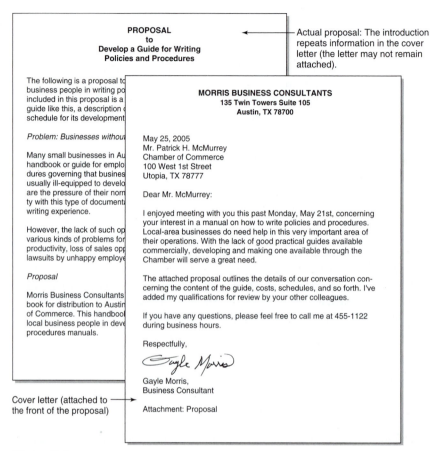

PROPOSAL
to
**Develop a Guide for Writing
Policies and Procedures**

◄——————Actual proposal: The introduction
repeats information in the cover
letter (the letter may not remain
attached).

The following is a proposal to
business people in writing po
included in this proposal is a
guide like this, a description o
schedule for its development

Problem: Businesses withou

Many small businesses in Au
handbook or guide for emplo
dures governing that busines
usually ill-equipped to develo
are the pressure of their norn
ty with this type of documenta
writing experience.

However, the lack of such op
various kinds of problems for
productivity, loss of sales opp
lawsuits by unhappy employe

Proposal

Morris Business Consultants
book for distribution to Austin
of Commerce. This handbool
local business people in deve
procedures manuals.

MORRIS BUSINESS CONSULTANTS
135 Twin Towers Suite 105
Austin, TX 78700

May 25, 2005
Mr. Patrick H. McMurrey
Chamber of Commerce
100 West 1st Street
Utopia, TX 78777

Dear Mr. McMurrey:

I enjoyed meeting with you this past Monday, May 21st, concerning
your interest in a manual on how to write policies and procedures.
Local-area businesses do need help in this very important area of
their operations. With the lack of good practical guides available
commercially, developing and making one available through the
Chamber will serve a great need.

The attached proposal outlines the details of our conversation concerning the content of the guide, costs, schedules, and so forth. I've
added my qualifications for review by your other colleagues.

If you have any questions, please feel free to call me at 455-1122
during business hours.

Respectfully,

Gayle Morris,
Business Consultant

Cover letter (attached to ——►
the front of the proposal)

Attachment: Proposal

Figure 5-6   Proposal with cover letter. The proposal proper uses a title at the top of the
page and repeats some of the contents of the cover letter (in case the letter is separated from
the proposal). (Example proposal is drawn from work done by Gayle Morris, student at
Austin Community College.)

envision it being passed around among reviewers, make it a separate document with its own title. Attach a cover memo to the front; in the memo,
restate the key elements of the introduction and the conclusion. (See Figure
5-6, but picture the letter reformatted as a memorandum.)

- *Separate proposal with cover letter.* If your proposal is long, if it's addressed
to people outside your company, and if you envision it being passed around
among reviewers, make it a separate document with its own title. Attach a
cover letter to the front; in the letter, restate the key elements of the introduction and the conclusion. (See Figure 5-6.)

PROGRESS REPORTS

Another common report type is variously called the progress report, status report, interim report, quarterly report, or monthly report. Its job is to present to your clients or associates the status of the work you are doing for them. They can then act as manager or executive of the project, modifying it or even canceling it if the need arises. In this situation, you are the supplier of the work of the project; the recipient of the work is the customer—the individual or organization that requires the work. Your client may be internal to your organization (for example, a work supervisor) or external (for example, a customer with whom you have a contract).

To understand the function of progress reports, think about the projects for which they are written. In most projects, changes, new and additional requirements, problems, and miscommunications are bound to occur. In this environment, clients may worry that the work is not being done properly, on schedule, or within budget. Suppliers, on the other hand, may worry that clients will not like how the project is developing, that new requirements jeopardize the schedule and budget for the project, or that unexpected problems affecting the schedule and outcome of the project have arisen. In this context, then, the progress report can allay clients' concerns about the schedule, quality, and cost of their projects; help suppliers stay in touch with their clients; maintain a professional image; and protect themselves from unreasonable expectations and mistaken or unwarranted accusations.

CONTENTS AND ORGANIZATION

Because of these functions and expectations, progress reports typically have the following contents and organization (see the excerpts from a progress report in Figure 5-7).

Introduction.   As with any report, start with the purpose and topic of the report, its intended audience, and a brief overview of the report's contents before diving into the thick of the discussion.

Project Description.   Briefly describe the project in case the progress report is routed to readers who are not familiar with the project. Summarize details such as the purpose and scope of the project, project start and completion dates, and names of suppliers and clients involved in the project. Unless the project changes, this description can become boilerplate text in future progress reports and appear under its own heading, enabling readers to skip it.

Progress Summary.   The real substance of the progress report is the discussion of what work you've completed, what work is in progress, and what's yet to come. This discussion can be handled several ways:

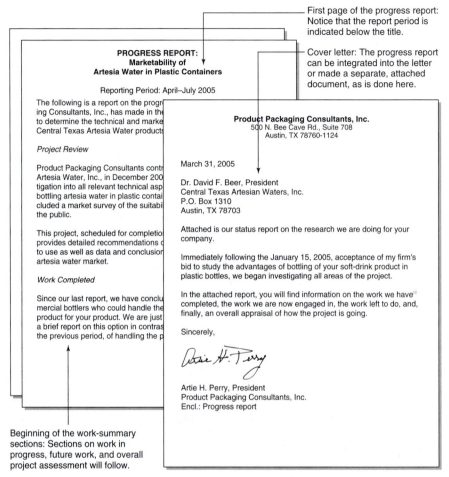

First page of the progress report: Notice that the report period is indicated below the title.

Cover letter: The progress report can be integrated into the letter or made a separate, attached document, as is done here.

PROGRESS REPORT:
Marketability of
Artesia Water in Plastic Containers

Reporting Period: April–July 2005

The following is a report on the progr
ing Consultants, Inc., has made in the
to determine the technical and marke
Central Texas Artesia Water products

Project Review

Product Packaging Consultants contr
Artesia Water, Inc., in December 200
tigation into all relevant technical asp
bottling artesia water in plastic contai
cluded a market survey of the suitabi
the public.

This project, scheduled for completio
provides detailed recommendations c
to use as well as data and conclusion
artesia water market.

Work Completed

Since our last report, we have conclu
mercial bottlers who could handle the
product for your product. We are just
a brief report on this option in contras
the previous period, of handling the p

Product Packaging Consultants, Inc.
500 N. Bee Cave Rd., Suite 708
Austin, TX 78760-1124

March 31, 2005

Dr. David F. Beer, President
Central Texas Artesian Waters, Inc.
P.O. Box 1310
Austin, TX 78703

Attached is our status report on the research we are doing for your company.

Immediately following the January 15, 2005, acceptance of my firm's bid to study the advantages of bottling of your soft-drink product in plastic bottles, we began investigating all areas of the project.

In the attached report, you will find information on the work we have completed, the work we are now engaged in, the work left to do, and, finally, an overall appraisal of how the project is going.

Sincerely,

Artie H. Perry, President
Product Packaging Consultants, Inc.
Encl.: Progress report

Beginning of the work-summary sections: Sections on work in progress, future work, and overall project assessment will follow.

Figure 5-7   Progress report—cover letter and first page. If your progress report is short, you can incorporate the report into a business letter (or memo), making it one continuous document (for an example of this approach, see the letter and memo proposals in Figure 5-5).

- *Time-periods approach.* Summarizes work completed in the previous period, work under way in the current period, and work planned for future periods.
- *Project-tasks approach.* Summarizes which tasks in the project have been completed, which tasks are currently under way, and which tasks are planned for future work.
- *Combined approach.* Combines these approaches by dividing the section on previous-period work into summaries of the work done on individual tasks. Or, this approach divides project-task sections into summaries of work completed, in progress, or planned for each task.

Use whichever of these approaches works best in terms of your project and the requirements of your client. For simpler projects, however, the time-periods approach works best. The project-tasks approach works well when the project has a number of semi-independent tasks on which you are working more or less concurrently.

Problems Encountered.   In this section, you go on record about the problems that have arisen in the project, problems you think may jeopardize the quality, cost, or schedule of the project.

Changes in Requirements.   In this section, you keep a history of changes in the project as you understand them: for example, schedule shifts, new requirements, and so on.

Overall Assessment of the Project.   In what is often the final section of the progress report, you give a general opinion as to how the project is going. In this section, resist the temptations to say that everything's going along just fine or to whine about every minor annoyance. Remember that your job is to provide your clients with the details they need to act as managers or executives of the project as a whole.

Other sections may also be required: for example, a summary of financial data on the project or the results of product testing. When you plan and write progress reports, be alert to the needs and expectations of your audience—in this case, those customers or supervisors on whom you depend for income or employment.

FREQUENCY OF PROGRESS REPORTS

If you're not sure whether progress reports are required for a project, especially short projects, check with your supervisor or client. Remember that the "progress report" may be nothing more than a quick email briefly describing the project status. It's a healthy impulse to avoid unnecessary work, but keep in mind that progress reports, when appropriate, strengthen your professional image. They keep you closer to your client, and they help eliminate unfortunate surprises.

The schedule for progress reports may be established by your supervisors or clients. If it's not, your sense of the project and the requirements of the client should dictate how many progress reports there should be and how often they should be delivered. Typically, progress reports are sent at the end of every month or every quarter. The larger the project, the more formally defined these requirements are and the more formal the progress reports are.

FORMAT

For large projects, progress reports can be lengthy, 100-page, bound, formal reports. Even so, the contents and organization are essentially the same as previously discussed. The formal elements include title page, table of contents, abstracts, and appendixes. (See Chapter 6 for details.) It's more likely, however, that your progress report can fit easily into a business letter or memorandum.

Note Lengthy? See the Department of Energy's *FY 2002 Progress Report for Hydrogen, Fuel Cells, and Infrastructure Technologies Program* at www.eere.energy.gov/hydrogenandfuelcells/annual_report.html. It's 600 pages!

INSTRUCTIONS

In your engineering career, you may often find yourself writing step-by-step procedures for employees, colleagues, customers, or clients. In such instructions, you explain how to assemble, operate, or troubleshoot some new product your team is working on or how to operate equipment around the office, laboratory, or site.

SOME PRELIMINARIES

Critical in instructions writing is putting yourself in your readers' place, making no unwarranted assumptions about their background or knowledge, and providing them everything they need to successfully complete the procedure.

Understand the important difference between instructions and product specifications. In rushed development cycles, product specifications are sometimes used, with little revision, as instructions. That's unfortunate: Specifications (discussed previously in this chapter) do not function well as instructions. Specifications approach a product as a group of features and functions—not in terms of tasks. Consider your microwave oven: The statement "Power Cook button enables user to set the power level" is a specification, not an instruction. The user also needs to know which other buttons to press and in what sequence. In specifications, the heading might be "Power Cook Function," whereas in instructions the heading would be something like "Cooking with Different Power Levels."

Critical in preparing to write instructions is audience analysis—identifying the relevant characteristics of the readers most likely to use your instructions. (For a full discussion of this task, see Chapter 2.) Consider what you expect your readers to know already and what you will explain in your instructions. For example, in explaining how to install a computer program, you have to decide whether readers understand some basics about installation media, folders, directories, and files.

CONTENTS AND ORGANIZATION

Introduction. In the introduction, include some combination of the following:

- *Subject.* Indicate the procedure you'll explain.
- *Product.* If you are providing instructions for a product, identify it.

- *Audience.* Indicate the knowledge or background your readers need in order to understand your instructions. If no special background is needed, indicate that as well.
- *Overview.* Briefly list the main contents of the instructions; for example, list the major tasks or procedures to be presented.

Special Notices. Most instructions contain specially formatted notices for warnings, cautions, and dangers. Often these appear in the introduction as well as in the body of the instructions at those points where they apply. If you neglect to include these special notices, you may find yourself in a lawsuit if readers injure themselves or lose money. (See Chapter 11 for more on this.)

Style and format of special notices vary widely, but here's a recommended approach:[2]

- *Note:* For emphasizing special points or exceptions that might otherwise be overlooked.
- *Attention:* For alerting readers to a potential for ruining the outcome of the procedure or damaging the equipment. (This is the "warning" in Figure 5-9.)
- *Caution:* For alerting readers to the possibility of minor injury because of some existing condition (for example, the hazard of paper cuts when opening a ream of paper). Also used when a potentially dangerous situation might develop because of some unsafe practice (for example, making an unapproved hardware modification).
- *Danger:* For calling attention to a situation that is potentially lethal or extremely hazardous to people (for example, exposed high-voltage wires as a result of removing a computer side panel). Use this notice with discretion, reserving it for situations where irreparable injury or loss of life could occur unless extreme care is used.

See Figures 5-8 and 5-9 for examples of these special notices. Serious ones are placed before the point at which readers might wreck their procedure, ruin their equipment, hurt somebody, or blow themselves up!

Background. For certain complex tasks, readers need to know conceptually what they are doing and why they are doing it. Positioned after the introduction and before the actual procedures, a background section enables readers to figure out much of the procedure, or its finer points, on their own. However, make sure that the background is closely related to the procedure. As a way of avoiding background with no immediate relation to the procedures it is supposed to elucidate, write background only *after* you've written the procedures.

---

[2] Many thanks to Linda St.Clair, editor, AIX & eServer pSeries Information Design & Development, IBM Corporation, for updates on notices.

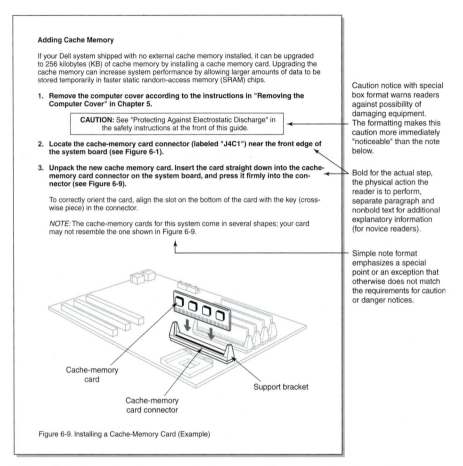

Figure 5-8 Instructions—excerpts. Notices, cross-references, and step formatting are shown. *Source:* Dell Computer Corporation. *Dell Dimension T Series Midsize Systems User's Guide* (adapted with permission).

Equipment and Supplies. List the supplies and equipment that readers must gather before they begin. Supplies are consumable items used in the process: paper, flour, glue, sandpaper, eggs, milk, nails, paint, paint thinner, sugar, and so forth. Equipment is the tools and machinery that are needed: screwdrivers, hammers, metering devices, and so on. For some instructions, it's not enough merely to list equipment and supplies. You must also specify sizes, brands, types, models, and so on.

Structure of the Instructions. Before you dive into the step-by-step discussion, identify the *tasks* in the procedure. Your instructions may have a simple series of steps that readers perform in sequence. For example, changing the oil in a car involves one task, a series of steps that must be performed in order—otherwise, you'll end up with oil all over the driveway, a burned-up engine, or both.

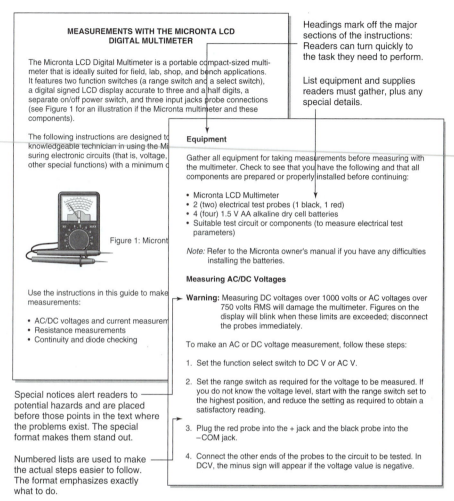

MEASUREMENTS WITH THE MICRONTA LCD DIGITAL MULTIMETER

The Micronta LCD Digital Multimeter is a portable compact-sized multimeter that is ideally suited for field, lab, shop, and bench applications. It features two function switches (a range switch and a select switch), a digital signed LCD display accurate to three and a half digits, a separate on/off power switch, and three input jacks probe connections (see Figure 1 for an illustration if the Micronta multimeter and these components).

The following instructions are designed to knowledgeable technician in using the Mi suring electronic circuits (that is, voltage, other special functions) with a minimum o

Figure 1: Micront

Use the instructions in this guide to make measurements:

- AC/DC voltages and current measurem
- Resistance measurements
- Continuity and diode checking

Headings mark off the major sections of the instructions: Readers can turn quickly to the task they need to perform.

List equipment and supplies readers must gather, plus any special details.

Equipment

Gather all equipment for taking measurements before measuring with the multimeter. Check to see that you have the following and that all components are prepared or properly installed before continuing:

- Micronta LCD Multimeter
- 2 (two) electrical test probes (1 black, 1 red)
- 4 (four) 1.5 V AA alkaline dry cell batteries
- Suitable test circuit or components (to measure electrical test parameters)

Note: Refer to the Micronta owner's manual if you have any difficulties installing the batteries.

Measuring AC/DC Voltages

Warning: Measuring DC voltages over 1000 volts or AC voltages over 750 volts RMS will damage the multimeter. Figures on the display will blink when these limits are exceeded; disconnect the probes immediately.

To make an AC or DC voltage measurement, follow these steps:

1. Set the function select switch to DC V or AC V.

2. Set the range switch as required for the voltage to be measured. If you do not know the voltage level, start with the range switch set to the highest position, and reduce the setting as required to obtain a satisfactory reading.

3. Plug the red probe into the + jack and the black probe into the –COM jack.

4. Connect the other ends of the probes to the circuit to be tested. In DCV, the minus sign will appear if the voltage value is negative.

Special notices alert readers to potential hazards and are placed before those points in the text where the problems exist. The special format makes them stand out.

Numbered lists are used to make the actual steps easier to follow. The format emphasizes exactly what to do.

Figure 5-9 Instructions—excerpts. Headings, lists, graphics, and special notices are critical elements of good instructions—but so is good, clear writing. (Example instructions are drawn from work done by Robert Hutchinson, student at Austin Community College.)

However, some instructions may describe several tasks that can be performed in practically any combination or order. Operating voice mail involves numerous tasks, some of which you perform only occasionally (recording a new greeting); others, every day (playing back messages or deleting messages). If that's the case, then use headings to help readers find these tasks quickly.

Discussion of the Steps. When you discuss the individual steps (the individual actions readers take to accomplish the procedure), be aware of some issues involving writing style, format, headings, and content:

- *Imperative writing style.* Many sentences use the imperative (for example, "Press the Enter key" or "Calculate the square footage in the lot") or are phrased with the word "you" (for example, "You should check the temperature of the . . ."). Don't hesitate to use this "in-your-face" style of writing; address readers directly, get their full attention, and be straightforward about what they are supposed to do.

- *Supplemental explanation.* Some individual steps may require additional explanation—for example, why readers should or should not do something. You may need to define potentially unfamiliar terms or describe how things look before, during, or after individual steps. Notice how the instructions in Figure 5-8 visually separate instructions from the supplemental explanation by bolding the actual instructions and skipping a line.

- *Special format.* When you explain the individual steps, use numbered lists for sequential steps. Use bulleted lists for steps in no necessary order (for example, nonsequential troubleshooting suggestions). The vertical-list format helps readers follow the procedure and visually cues them for each specific action to perform.

- *Headings.* For all but the simplest instructions, use headings. Headings enable readers to find supply lists, background information, and troubleshooting tips. Headings guide readers to specific tasks. For example, using well-designed headings in voice-mail instructions, readers can quickly find the section on how to forward messages.

GRAPHICS IN INSTRUCTIONS

Graphics are usually essential in instructions. Sometimes words cannot convey enough detail about key objects and key actions in a procedure. For example, just how does part A fit into part B? Make a list of the *key objects* and the *key actions* in your instructions, and identify those that readers might have trouble with. These are the ones for which you may need graphics (see Chapter 7).

RECOMMENDATION REPORTS

A recommendation report evaluates or promotes an idea—for example, an endorsement of telecommuting for fellow employees. The context can vary: Management might direct you to study the feasibility of telecommuting and to make recommendations, or management might direct you to compare telecommuting products and to recommend one. The common elements are recommendations and, as the following discussion shows, a comparative discussion that supports those recommendations.

Where you work, it may be called a recommendation report, an evaluation report, a feasibility report—or even a proposal. But the essential structure is the same for all—comparing options and recommending one.

SOME DISTINCTIONS

A recommendation report, as its name indicates, makes a recommendation about plans, products, or people. In its simplest form, it establishes certain requirements (often called criteria), compares two or more options, and recommends one. Other elements may be involved: for example, background on the technology; descriptions of the options; an explanation of how the field was narrowed; even discussion of the technical, economical, and social practicality of the idea.

Typically, the terms *proposal, feasibility report, evaluation report,* and *recommendation report* are used interchangeably. Don't expect much precision in real-world usage of these terms. In this book, we make the following distinctions:

- *Recommendation reports:* Compare two or more options against each other (and against certain requirements) and then make a recommendation.
- *Evaluation reports:* Compare an idea, program, or thing against criteria or requirements as a means of determining its value. This type may recommend, but essential is the statement of the value of the idea, program, or thing.
- *Feasibility reports:* Compare a project against requirements relating to its economic, technical, or social practicality, and then recommend whether the project should be initiated.
- *Proposals:* Make a bid or seek approval to do a project and then supply supporting information on the proposer's qualifications. The primary task is to land a contract or get approval. (See the section on proposals earlier in this chapter.)

Each of these types works toward an endorsement, recommendation, or value judgment; your job as the writer is to achieve that end.

CONTENTS AND ORGANIZATION

To write this type of report, remember that you must provide data and conclusions so that readers can decide for themselves whether your recommendations are justified.

Introduction. As with any introduction, indicate the purpose of the document. Indicate right up front that the purpose is to recommend something for some specific use or situation. Indicate the audience—the intended readers of the report and any technical background they need (and if none is needed, say so). Also include an overview: Provide a brief list of the contents of the report.

Background on the Situation. Consider whether you should discuss the situation in which the recommendation report is needed. The immediate audience may know perfectly well what the situation is, but your report may get passed around to others who don't. Background may also prove a helpful memory jogger for some overly busy readers. (Remember, headings enable readers who know this background to skip over it.)

Requirements. In practically any recommendation, there are requirements such as cost, operational features, size, flow rates, weight, and so on. Consider the example of selecting email software for telecommuting: What are the specifications? Ease of use? Versions for Macintosh and PC machines? File transfer capability? Bulletin board features? In your recommendation *study*, you determine these requirements. In your recommendation *report*, you describe these requirements. Readers can then consider these requirements and decide for themselves whether they agree.

Technical Background. For some recommendation reports, it may be necessary to provide some brief technical discussion. In the early days of CD-ROM products, you might have discussed 8- and 16-bit technology, triple or quadruple spin, sampling, and other related technical concepts. As with most introductory and background sections, it's best to write them later in your process. Write the heart of the recommendation report first—the comparisons, conclusions, and recommendations.

Description. In some recommendation reports, you may need to describe the options you are comparing. These descriptions are neutral—no comparisons or evaluations are provided. For example, in a recommendation on laser printers, describe each of the finalists separately in terms of its size, dimensions, operating features, warranties, upgrade possibilities, and so on. (See Figure 5-10 for an example.)

Point-by-Point Comparisons. Comparisons constitute one of the three main sections of a recommendation report. Comparison sections focus on specific points, such as cost, ease of use, warranties, and so on. For example, in a section on cost, the cost of each option is compared. Usually, it's not a simple matter of one being the cheapest and another being the most expensive. Things get blurred by special features and service plans that can be added on. In these cases, help readers: Untangle the complexities for them and point to the best choice.

End each comparative section with a statement as to which option is best in terms of that comparative category. In Figure 5-11, notice that the comparison of power performance ends with a conclusion as to which option provides the best performance; the emissions section ends with a conclusion as to which option is best in terms of low emissions.

Remember: You write these comparisons so that readers can see your logic—how you reached your conclusions. Give readers a chance to disagree with your thinking and to reach their own conclusions and recommendations.

Background on Hybrid Electronic Vehicles

The hybrid electronic vehicle (HEV) uses a gasoline-electric powertrain that creates an environmentally conscious, fuel-efficient hybrid. The powertrain consists of a gasoline engine, an electric motor, and an energy storage device. The battery is automatically recharged during braking and deceleration and does n[...] result, fuel efficiency is greatly increa[...] fuel than vehicles powered by gasoli[...] decreased. Also, HEV engines can b[...] driver's average load, not peak load,[...] increasing fuel efficiency.

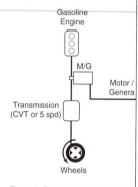

Gasoline
Engine

M/G

Motor /
Genera[...]

Transmission
(CVT or 5 spd)

Wheels

Figure 1. Component placement in th[...]
"M/G" charges the battery from the g[...]
the wheels, charges the battery from[...]
starts or stops the gasoline engine [5[...]

Descriptions

As of 2002, three HEVs were prominently available for purchase in the United States. They were the Honda Insight, Honda Civic, and the Toyota Prius (referred to as the Civic, the Insight, and the Prius, respectively). The cost of all three is right at $20,000; all three come equipped with air conditioning, power windows, stereo, and other similar amenities.

Figure 2. Honda Insight, Toyota Prius, Honda Civic Hybrid (2002 models) [4].

Honda Insight. The Honda Insight is a 2-passenger vehicle with a 1.0-liter 3-cylinder engine. The engine and motor have combined output of 67 hp at 5,700 rpm, 66 lb-ft of torque at 4,800 rpm. Honda claims a fuel efficiency of 57 mpg city and 56 mpg highway (automatic transmission). Weighing 1,878 pounds, the vehicle accelerates from 0 to 60 mph in 11.3 seconds (manual transmission) and 11.2 (automatic transmission). Its electric motor employs a permanent magnet with peak power of 13 hp at 3,000 rpm and 36 lb-ft of torque. It uses a sealed nickel-metal hydrid (NiMH) battery that produces 144 volts [6, 11].

Honda Civic Hybrid. The Honda Civic Hybrid is a 5-passenger vehicle with a 1.3-liter 4-cylinder engine. The engine and motor combined put out 85 hp at 5,700 rpm, 87 lb-ft of torque at 3,300 rpm. Honda claims a fuel efficiency of 47 mpg city and 48 mpg highway (automatic). Weighing 2,740 pounds, the vehicle accelerates from 0 to 60 mph in 11.2 (automatic). Its electric motor employs a permanent magnet with peak power of 13.4 hp at 4,000 rpm. It uses a sealed nickel-metal hydrid (NiMH) battery that produces 144 volts [2, 6].

Figure 5-10   Recommendation report—technology and description sections. Because the hybrid technology is new, the writer discusses it, followed by descriptions of the three vehicles to be compared. (Be aware that the report begins with an introduction and requirement section.) (Example is adapted from a report by Cynthia Hale, technical writing student, Ohio University.)

Conclusions (Summary).   The conclusions section is a repeat of the conclusions you reached in each of those individual comparative sections. Notice that items 2 and 4 in the section entitled "Summary" shown in Figure 5-12 echo the conclusions stated in Figure 5-11.

In some cases, no individual option may prove to be the obvious, best choice. One option may be the cheapest; another may be the most reliable; another may be the easiest to use; still another may have far more functions and features. These are the *primary conclusions*. But how do you pick a "winner" when they conflict? If you've defined them carefully, your requirements should point to the final recom-

Comparison Points

Recommendations in this report will be based on the following criteria:
(1) fuel efficiency, (2) power performan

Fuel efficiency. The Honda Insight cla
city and 56 mpg highway [11]. The Ho
efficiency of 35 mpg city and 44 mpg h
fuel efficiency of 52 mpg city and 45 m
vehicles with automatic transmission.)

Although the Insight has the best fuel
compared, its lack of certain safety fea
less desirable. The fact that the Prius
efficiency, together with its other asset
choice in this category.

Power performance. The performanc
primarily by power outputs, acceleratio
engine, electric motor, and battery. Ta
The Insight is equipped with a 1.0-liter
Civic Hybrid is equipped with a larger,
Prius is equipped with an even larger

The Insight's gasoline engine and elec
hp at 5700 rpm and 66 lb-ft of torque a
output of 85 hp at 5700 rpm and 87 lb-
Prius, a combined output of 70 hp at 4
rpm [7].

The Insight's electric motor employs a
13 hp at 3000 rpm and 36 lb-ft of torqu
same, at 13.4 hp at 4000 rpm, while th
most powerful, producing 44 hp at 560
rpm.

The Insight accelerates from 0 to 60 mph in 11.2 seconds; the Civic in 11.2 seconds; and the Prius in 12.8 seconds.

Table 1. Power Performance

Category	Prius	Civic	Insight
Engine size	1.5 liter, 4 cyl.	1.3 liter, 4 cyl.	1.0 liter, 3 cyl.
Combined power			
HP (hp @ rpm)	70 @ 4500	85 @ 5700	67 @ 5700
Torque (lb-ft at rpm)	82 @ 4200	87 @ 3300	66 @ 4800
Electric motor power			
HP (hp @ rpm)	44 @ 5600	13.4 @ 4000	13 @ 3000
Torque (lb-ft at rpm)	258 @ 500	unavailable	36 @ 3000
Battery (volts)	274	144	144
Acceleration (0–60 mph/sec)	12.8	11.2	11.2

Based on these comparisons, the Civic provides better overall power performance than both the Prius and the Insight.

Emissions and efficiency. Table 2 provides emissions and efficiency figures for the Prius, the Civic, and the Insight.

Table 2. Emission and Efficiency Ratings [10]

Category	Prius	Civic	Insight
Energy efficiency (city/hwy mpg)	52/45	35/44	57/56
EPA smog-forming emissions rating*	6.5	7	10
Greenhouse gas emissions (tons/year)	4.0	4.9–5.2	3.1

*A score of 10 is best.

The overall emissions rating of the Insight is far and away the best, with the Prius coming in well ahead of the Civic.

Safety. The Civic, Insight, and Prius all have the following safety feat-

Individual comparisons: These occur at the end of each comparative section.

Figure 5-11   Recommendation report—comparison sections. The three vehicles are compared on each separate point, with a conclusion at the end of each section as to which vehicle rates best on that individual comparative point.

mendation. Requirements enable you to state secondary conclusions: conclusions that resolve conflicting primary conclusions. Notice how the seventh conclusion—a *secondary conclusion*—in Figure 5-12 weighs four different primary conclusions against each other:

Although the Insight has the best fuel efficiency, lowest cost, and lowest emissions, its light weight and safety ratings make it questionable as the best choice.

Recommendations.   The recommendations section simply states what has probably become obvious—which option is recommended. The example in Figure 5-12

Recommendations

The preceding comparisons suggest the following recommendations:
- Insight—if your primary concern is emissions control above all other factors.
  or
- Civic—if your priorities are performance and safety features.
  or
- Prius—if you want a balance of low emissions, performance, and safety.

Sources Cited

1. edmunds.com. "What's New for 2002." http://www.edmunds.com/new/2002/toyota/prius/. Accessed July 8, 2002.
2. hondacars.com. "Specifications: 2003 Civic Hybrid." http://www.hondacars.com/models/specifications.asp?ModelName=
3. hondacars.com. "Specifications: 2003 In models/specifications.asp?ModelName=
4. Insight/Central.net. "Comparison to Othe KB/compare/compare-honda-insight.htm
5. InsightCentral.net. "Hybrid Powertrain." compare/prius-powertrain.html. Accesse
6. InsightCentral.net. "Insight vs. Hybrid Ci compare/compare-civic.html. Accessed
7. InsightCentral.net. "Insight vs. Prius." ww prius.html. Accessed July 3, 2003.
8. toyota.com. "Prius Technology." www.to technology/prius technology.html. Acces
9. toyota.com. "Toyota Hybrid System." ww prius/technology/prius ths.html. Accesse
10. U.S. Department of Energy & Environme Emissions." www.fueleconomy.gov/feg/e July 3, 2003.
11. U.S. Department of Energy. "Technolog afdc.doe.gov/pdfs/insight snapshot.pdf.
12. U.S. Department of Energy. "Technolog ccities.doe.gov/pdfs/snapshot.pdf. Acce
13. U.S. Department of Energy. "What is an what.html. Accessed July 3, 2003.
14. U.S. Department of Energy. "Where Car hev/where.html. Accessed July 3, 2003.

Summary

The following summary was compiled from the comparisons of the Honda Insight, the Toyota Prius, and the Honda Civic Hybrid:

1. The Insight rates highest in fuel efficiency.
2. The Civic provides better power performance than the Prius and the Insight.
3. The Civic and the Insight provide somewhat better acceleration than the Prius.
4. The Insight has the best emissions rating of the three hybrids compared.
5. All three vehicles have minimum safety features, but the Civic has additional ones making it the safest car to drive.
6. All three vehicles are in the same price range, with the Insight being the least expensive.
7. Although the Insight has the best fuel efficiency, lowest cost, and lowest emissions, its light weight and safety ratings make it questionable as the best choice.
8. The fuel efficiency of the Civic make it questionable as the best choice.
9. The power performance of the Prius make it questionable as the best choice.

Table 2. Comparison Points for the Civic, Prius, and Insight

Category	Honda Civic	Toyota Prius	Honda Insight
Fuel efficiency	2	3	4
Power performance	3	2	2
Safety	4	3	2
Cost	3	3	4
Emissions	3	3	4
Total	15	14	16

*1 = Poor, 2 = Fair, 3 = Good, 4 = Excellent

Figure 5-12   Recommendation report—conclusions and recommendation sections. Notice that the numbered statements in the conclusions section ("Summary") repeat the conclusions from the individual comparative sections earlier in the report. Notice too that the summary table provides an alternate presentation of the numbered list of conclusions.

briefly mentions which comparisons were most influential in reaching the final recommendation.

Sometimes, recommendations may not be so obvious either. To make recommendations, you may have to state qualifications. In Figure 5-12, for example:

> • If your primary concern is emissions control above all other factors, choose the Insight.
> • If your priorities are performance and safety features, choose the Civic.
> • If you want a balance of low emissions, performance, and safety, choose the Prius.

Sometimes, you may not be able to recommend *any* of the options. Imagine a study on grammar-checking software. You investigate as many of the different applications as you can, read the reviews, and get as many product demos as you can. At the end, you throw up your hands, finding them all worthless, and recommend that your company would be better off to hire a human being to do the copyediting.

GRAPHICS IN RECOMMENDATION REPORTS

As you plan a recommendation report, consider the illustrations, drawings, tables, or charts that might be necessary. In this type of report, tables are often effective, as is shown in Figures 5-11 and 5-12. For more dramatic demonstrations of your points, you can use line graphs, pie charts, bar charts, and other such ways of depicting data. (For more on graphics and tables, see Chapter 7.)

EXERCISES

Talk to several professional engineers about the reports they write, and ask them questions like the following:

1. Which of these types of reports do they most commonly write? Are there other types, not covered in this chapter, that they also write?

2. What are the chief purposes of the reports they write? Are the reports for internal or external consumption, for colleagues or clients?

3. How important are these reports to their business and professional careers? How much, for example, do they rely on proposals to get contracts? How often is their entire work product a written document?

4. Do they get editorial or production assistance in preparing these reports? Or do they handle the entire writing, editing, printing, and binding of their reports themselves?

5. Ask your engineering interviewees about the progress reports they write. What sorts of projects require progress reports? How often do they submit progress reports for a typical project?

BIBLIOGRAPHY

American National Standards Institute. *American National Standards for Product Safety Signs and Labels.* ANSI Z535 (1998). American National Standards Institute, 11 West 42nd Street, New York, NY 10036.

Bowman, J. P., and B. P. Branchaw. *How to Write Proposals That Produce.* Phoenix, AZ: Oryx, 1992.

Caher, John M. "Technical Documentation and Legal Liability." *Journal of Technical Writing and Communication* 25 (1995), 5–10.

Cavin, Janis I. *Understanding the Federal Proposal Review Process.* Washington, DC: American Association of State Colleges and Universities, 1984.

Clement, David E. "Human Factors, Instructions, Warnings, and Product Liability." *IEEE Transactions on Professional Communication* 30, no. 3 (1987), 149–156.

Dolphin, W. D. "Writing Lab Reports and Scientific Reports." www.mhhe.com/biosci/genbio/maderinquiry/writing.html. Accessed July 8, 2003.

Fischer, Martin A. *Engineering Specifications Writing Guide.* Englewood Cliffs, NJ: Prentice-Hall, 1983.

Heylar, P. S. "Products Liability: Meeting Legal Standards for Adequate Instructions." *Journal of Technical Writing and Communications* 22, no. 2 (1992), 125–147.

Hill, James W., and Timothy Wahlen. *How to Create and Present Successful Government Proposals.* New York: IEEE Press, 1993.

Miller, George A. "The Magical Number Seven, Plus or Minus Two: Some Limits on Our Capacity for Processing Information." In *The Psychology of Communication: Seven Essays.* New York: Basic Books, 1967.

Purdy, David C. *A Guide to Writing Successful Engineering Specifications.* New York: McGraw-Hill, 1990.

Stewart, Rodney D., and Anne L. Stewart. *Proposal Preparation.* New York: Wiley, 1992.

Velotta, Christopher. "Safety Labels: What to Put in Them, How to Write Them, and Where to Put Them." *IEEE Transactions on Professional Communication* 30 (1987), 121–136.

Wogalter, Michael S. "Factors Influencing the Effectiveness of Warnings." *Proceedings of Public Graphics,* 5.1–5.21. Department of Psychonomics, University of Utrecht, and Department of Industrial Design Engineering, Delft University of Technology, The Netherlands. www.electromark.com/help_hints/research/research_reg_specs.asp. Accessed July 8, 2003.

6

WRITING AN ENGINEERING REPORT

Reports are perhaps the most common documents that you will write both as engineering students and as engineers. Consequently, your success—both in school and in the workplace—will partly depend on your ability to produce effective reports. Unfortunately, learning to produce the various types of reports required is not always a simple matter. However, certain conventions provide guidelines for organization and content.

> Susan Stevenson and Steve Whitmore, *Strategies for Engineering Communication* (New York: John Wiley & Sons, 2002)

This report, by its very length, defends itself against the risk of being read.

> Winston Churchill

Engineers often get involved in projects that include writing reports. Engineering reports have specifications just like any other kind of project. Specifications for reports involve layout, organization and content, the format of headings and lists, the design of the graphics, and so on. In fact, the American National Standards Institute (ANSI) has published specifications for engineering reports entitled *Scientific and Technical Reports: Organization, Preparation, and Production.*

The advantage of a required structure and format for reports is that you or anyone else can expect them to be designed in a familiar way—you know what to look for and where to look for it. Reports are usually read in a hurry—people are in a hurry to get to the information they need, the key facts, the conclusions, and other essentials. A standard report format is like a familiar neighborhood.

When you analyze the design of an engineering report, notice how repetitive some sections are. This duplication has to do with how people read reports. They don't read reports straight through: They may start with the executive summary, skip around, and probably not read every page. Your challenge is to design reports so that these readers encounter your key facts and conclusions no matter how much of the report they read or in what order they read it.

The standard components of the typical engineering report are as follows:

Transmittal letter

Covers and label

Table of contents

List of figures

Executive summary

Introduction

Body of the report

Appendixes (including references)

The following sections guide you through each of these sections, pointing out the key features. As you read and use these guidelines, remember that these are *guidelines*, not commandments. Different companies, professions, and organizations have their own varied guidelines for reports—you'll need to adapt your practice to those as well as to the ones presented here.

LETTER OF TRANSMITTAL

The transmittal letter is a cover letter. An example is shown in Figure 6-1. It is either attached to the outside of the report with a paper clip or is bound within the report as part of the historical record of that report. It is a communication from you—the report writer—to the recipient, the person who requested the report and who may even be paying you for your expert consultation. Essentially, it says, "Okay, here's the report that we agreed I'd complete by such-and-such a date. Briefly, it contains this and that, but does not cover this or that. Let me know if it meets your needs." The transmittal letter explains the context—the events that brought about the report. It contains information about the report that does not belong in the report.

In Figure 6-1, notice the standard business-letter format. If you write an internal report, use the memorandum format instead; in either case, the contents and organization are the same:

- *First paragraph:* Cites the name of the report, putting it in italics. It also mentions the date of the agreement to write the report.
- *Middle paragraph:* Focuses on the purpose of the report and gives a brief overview of the report's contents.

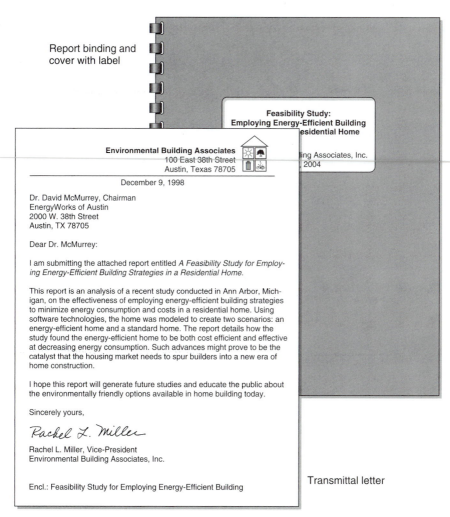

Report binding and cover with label

Feasibility Study:
Employing Energy-Efficient Building
...esidential Home

...ing Associates, Inc.
. 2004

Environmental Building Associates
100 East 38th Street
Austin, Texas 78705

December 9, 1998

Dr. David McMurrey, Chairman
EnergyWorks of Austin
2000 W. 38th Street
Austin, TX 78705

Dear Dr. McMurrey:

I am submitting the attached report entitled *A Feasibility Study for Employ-ing Energy-Efficient Building Strategies in a Residential Home.*

This report is an analysis of a recent study conducted in Ann Arbor, Mich-igan, on the effectiveness of employing energy-efficient building strategies to minimize energy consumption and costs in a residential home. Using software technologies, the home was modeled to create two scenarios: an energy-efficient home and a standard home. The report details how the study found the energy-efficient home to be both cost efficient and effective at decreasing energy consumption. Such advances might prove to be the catalyst that the housing market needs to spur builders into a new era of home construction.

I hope this report will generate future studies and educate the public about the environmentally friendly options available in home building today.

Sincerely yours,

Rachel L. Miller

Rachel L. Miller, Vice-President
Environmental Building Associates, Inc.

Encl.: Feasibility Study for Employing Energy-Efficient Building

Transmittal letter

Figure 6-1 Transmittal letter and report front cover. Normally the transmittal letter is paperclipped to the front of the report. (Logo was borrowed with permission from green-builder.com; example material was adapted from a report by Rachel L. Miller, engineering student, University of Texas at Austin.)

- *Final paragraph:* Encourages the reader to get in touch if there are questions, comments, or concerns. It closes with a gesture of good will, expressing hope that the reader finds the report satisfactory.

As with any other element in an engineering report, you may have to modify the contents of this letter for specific situations. For example, you might want to add another paragraph, listing questions you'd like readers to consider as they review the report.

COVER AND LABEL

If your report is over ten pages, bind it in some way and create a label for the cover.

COVERS

Covers give reports a solid, professional look as well as protection. You can choose from many types of covers. When you go to the copy shop, keep these tips in mind:

- Totally unacceptable are the clear (or colored) plastic slip cases with the plastic sleeve on the left edge. These are like something out of freshman English; plus, they are aggravating to use—readers must struggle to keep them open and hassle with the static electricity they generate.

- Marginally acceptable are the covers for which you punch holes in the pages, load the pages, and bend down the brads. If you use this type, leave an extra half-inch margin on the left edge so that readers don't have to pry the pages apart. Of course, this type of cover prevents pages from lying flat: Readers must grab available objects or use various body parts to keep the pages weighted down.

- By far the best covers are those that allow reports to lie open by themselves. What a great relief for a report to lie open in your lap or on your desk. This type uses a plastic spiral for the binding and thick, card-stock paper for the covers. Check with your local copy shop for these types of bindings; they are inexpensive and add to the professionalism of your work.

- Generally less preferable are loose-leaf notebooks or ring binders. These are too bulky for short reports, and the page holes tend to tear. Of course, the ring binder makes changing pages easy; if that's how your report will be used, then it's a good choice.

- At the "high end" are the overly fancy covers with their leatherette look and gold-colored trim. Avoid them—keep the cover plain, simple, and functional.

LABELS

Be sure to devise a label for the cover of your report. It's a step that some report writers forget. Without a label, a report is anonymous; it gets ignored.

The best way to create a label is to use your word processing software to design one on a standard page with a graphic box around the label information. Print it out, then go to a copy shop and have it photocopied directly onto the report cover.

Not much goes on the label: the report title, your name, your organization's name, a report tracking number, and a date. There are no standard requirements for the label, although there should be in your company or organization. (An example of a report label is shown in Figure 6-1.)

PAGE NUMBERING

The style for numbering pages in a report can be summarized by the following rules:

- All pages in the report (within but excluding the front and back covers) are numbered; however, on some pages, the numbers are not displayed.
- In the contemporary design style, all pages throughout the document use arabic numerals; in the traditional design style, all pages *before* the introduction (the first page of the body of the report) use lowercase roman numerals.
- On special pages, such as the title page and the first page of the introduction, page numbers are not displayed.
- Page numbers can be placed in one of several areas on the page. Usually, the best and easiest choice is to place page numbers at the bottom center of the page (remember to hide them on special pages).
- If you place page numbers at the top of the page, you must hide them on chapter or section openers where a heading or title is at the top of the page.

Note   Longer reports often use the page-numbering style known as folio-by-chapter or double-enumeration (for example, pages in Chapter 2 would be numbered 2-1, 2-2, 2-3, and so on). This style eases the process of adding and deleting pages.

ABSTRACT AND EXECUTIVE SUMMARY

Most engineering reports contain at least one abstract—sometimes two, in which case the abstracts play different roles. Abstracts summarize the contents of a report, but the different types do so in different ways:

- *Descriptive abstract.* This type provides an overview of the purpose and contents of the report. In some report designs, the descriptive abstract is placed at the bottom of the title page.
- *Executive summary.* Another common type is the executive summary, which also summarizes the key facts and conclusions contained in the report. (See the example shown in Figure 6-2.) It's as if you used a yellow highlighter to mark the key sentences in the report and then siphoned them all out onto a separate page and edited them for readability. Typically, executive summaries are one-tenth to one-twentieth the length of reports 10 to 50 pages long. For

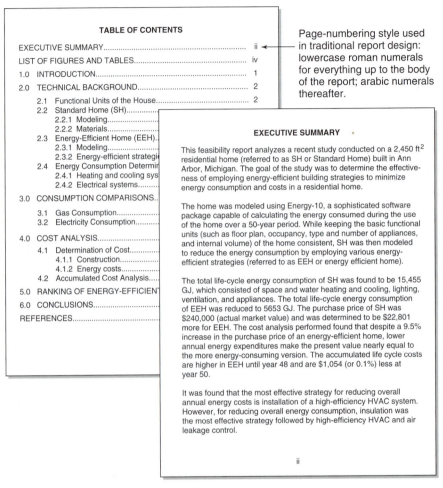

Figure 6-2   Title page and table of contents from an engineering report. (Some reports include a descriptive abstract at the top or bottom of the title page.)

longer reports, ones over 50 pages, the executive summary should not go over 3 typewritten pages. The point of the executive summary is to provide a summary of the report—something that can be read quickly.

If the executive summary, introduction, and transmittal letter strike you as repetitive, remember that readers don't necessarily start at the beginning of a report and read page by page to the end. They skip around: They may scan the table of contents; they usually skim the executive summary for key facts and conclusions. They may read carefully only a section or two from the body of the report, and then skip the rest. For these reasons, reports are designed with some duplication so that readers will be sure to see the important information no matter where they dip into the report.

TABLE OF CONTENTS

You're familiar with tables of contents (TOCs) but may never have stopped to look at their design. The TOC shows readers what topics are covered in the report, how those topics are discussed (the subtopics), and on which page numbers those sections and subsections start.

In creating a TOC, you have a number of design decisions:

- *How many levels of headings to include.* In longer reports, consider *not* including all of the lower-level headings in order to keep the TOC from becoming long and unwieldy. The TOC should provide an at-a-glance way of finding information in the report quickly.

- *Indentation, spacing, and capitalization.* Notice in Figure 6-2 that each of the three levels of headings are aligned with each other. Although you can't see it in Figure 6-2, page numbers are right-aligned with each other. Notice also the capitalization: Main chapters or sections are all caps; first-level headings use initial caps on each main word; lower-level sections use initial caps on the first word only.

- *Vertical spacing.* Notice that the first-level sections have extra space above and below, which increases readability.

One final note: Make sure the words in the TOC are the same as they are in the text. As you write and revise, you might change some of the headings—don't forget to change the TOC accordingly.

LISTS OF FIGURES AND TABLES

The list of figures and list of tables have many of the same design considerations as the table of contents. Readers use these lists to find the illustrations, diagrams, tables, and charts in your report.

Complications arise when you have both tables and figures. Strictly speaking, *figures* are illustrations, drawings, photographs, graphs, and charts. *Tables* are rows and columns of words and numbers; they are not considered figures.

For longer reports that contain dozens of figures and tables each, create separate lists of figures and tables. Put them together on the same page if they fit, as shown in Figure 6-3. You can combine the two lists under the heading "List of Figures and Tables," as shown in the TOC in Figure 6-2.

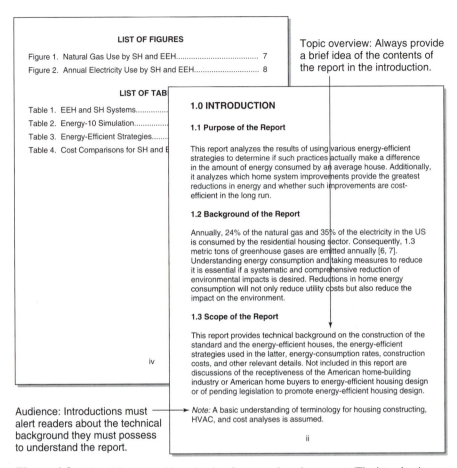

Topic overview: Always provide a brief idea of the contents of the report in the introduction.

LIST OF FIGURES

LIST OF TAB

1.0 INTRODUCTION

1.1 Purpose of the Report

This report analyzes the results of using various energy-efficient strategies to determine if such practices actually make a difference in the amount of energy consumed by an average house. Additionally, it analyzes which home system improvements provide the greatest reductions in energy and whether such improvements are cost-efficient in the long run.

1.2 Background of the Report

Annually, 24% of the natural gas and 35% of the electricity in the US is consumed by the residential housing sector. Consequently, 1.3 metric tons of greenhouse gases are emitted annually [6, 7]. Understanding energy consumption and taking measures to reduce it is essential if a systematic and comprehensive reduction of environmental impacts is desired. Reductions in home energy consumption will not only reduce utility costs but also reduce the impact on the environment.

1.3 Scope of the Report

This report provides technical background on the construction of the standard and the energy-efficient houses, the energy-efficient strategies used in the latter, energy-consumption rates, construction costs, and other relevant details. Not included in this report are discussions of the receptiveness of the American home-building industry or American home buyers to energy-efficient housing design or of pending legislation to promote energy-efficient housing design.

iv

Audience: Introductions must alert readers about the technical background they must possess to understand the report.

Note: A basic understanding of terminology for housing constructing, HVAC, and cost analyses is assumed.

ii

Figure 6-3 List of figures and introduction for an engineering report. (The introduction comes *after* the list of figures.)

INTRODUCTION

An essential element of any report is its introduction—make sure you are clear on its real purpose and contents. In an engineering report, the introduction prepares the reader to read the main body of the report. It does not dive into the subject, although it may provide a bit of theoretical or historical background. Instead, introductions indicate or discuss the following (not necessarily in this order):

- Specific purpose and topic of the report (indicated somewhere in the first paragraph)
- Intended audience of the report—knowledge or experience that readers need in order to understand the report

- Situation that brought about the need for the report
- Scope of the report—topics covered as well as topics not covered (specifically, ones that some readers might expect)
- Background (such as concepts, definitions, history, statistics)—just enough to get readers interested and to enable them to understand the context

Review the introduction in Figure 6-3 to see how these elements are handled.

The introduction is often mistakenly considered to be synonymous with background information. As the preceding list shows, background is only a minimal part of an introduction. Remember: The introduction prepares readers to read the report; it "introduces" them to the report. If the background gets out of hand and runs on for too many pages, move it to a section of its own, either just after the introduction or in an appendix. For a typical 20-page report, for example, the introduction shouldn't be too long—no more than 2 pages—and the background within the introduction should only be about a third of the introduction.

BODY OF THE REPORT

The body of the report is of course the main text of the report, the sections between the introduction and conclusion. Figures 6-4 and 6-5 show a couple of sample pages.

HEADINGS

In all but the shortest reports (two pages or less), use headings to mark off the different topics and subtopics covered. Headings enable readers to skim your report and dip down at those points where you present information that they want. Notice that the headings in the examples throughout this chapter use the decimal style (see Chapter 2 for discussion).

LISTS

In the body of a report, also use bulleted, numbered, and two-column lists where appropriate. Lists help by emphasizing key points, by making information easier to follow, and by breaking up solid blocks of text. For example, if you have three key points that readers must not overlook, use a bulleted list. If you have a sequence of steps that readers must perform, use a numbered list. If you have key terms and definitions that need to stand out, use a two-column list. (See Chapter 2 for discussion and guidelines.)

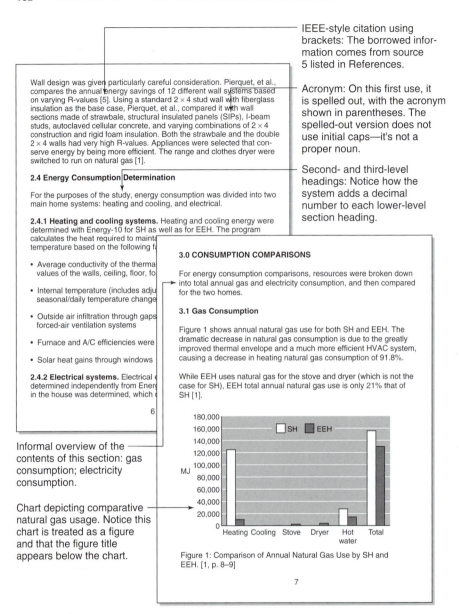

IEEE-style citation using brackets: The borrowed information comes from source 5 listed in References.

Wall design was given particularly careful consideration. Pierquet, et al., compares the annual energy savings of 12 different wall systems based on varying R-values [5]. Using a standard 2 × 4 stud wall with fiberglass insulation as the base case, Pierquet, et al., compared it with wall sections made of strawbale, structural insulated panels (SIPs), I-beam studs, autoclaved cellular concrete, and varying combinations of 2 × 4 construction and rigid foam insulation. Both the strawbale and the double 2 × 4 walls had very high R-values. Appliances were selected that conserve energy by being more efficient. The range and clothes dryer were switched to run on natural gas [1].

Acronym: On this first use, it is spelled out, with the acronym shown in parentheses. The spelled-out version does not use initial caps—it's not a proper noun.

2.4 Energy Consumption Determination

Second- and third-level headings: Notice how the system adds a decimal number to each lower-level section heading.

For the purposes of the study, energy consumption was divided into two main home systems: heating and cooling, and electrical.

2.4.1 Heating and cooling systems. Heating and cooling energy were determined with Energy-10 for SH as well as for EEH. The program calculates the heat required to maint... temperature based on the following f...

3.0 CONSUMPTION COMPARISONS

For energy consumption comparisons, resources were broken down into total annual gas and electricity consumption, and then compared for the two homes.

- Average conductivity of the thermal values of the walls, ceiling, floor, fo...

- Internal temperature (includes adju... seasonal/daily temperature change...

- Outside air infiltration through gaps... forced-air ventilation systems

- Furnace and A/C efficiencies were...

- Solar heat gains through windows

2.4.2 Electrical systems. Electrical... determined independently from Ener... in the house was determined, which...

3.1 Gas Consumption

Figure 1 shows annual natural gas use for both SH and EEH. The dramatic decrease in natural gas consumption is due to the greatly improved thermal envelope and a much more efficient HVAC system, causing a decrease in heating natural gas consumption of 91.8%.

While EEH uses natural gas for the stove and dryer (which is not the case for SH), EEH total annual natural gas use is only 21% that of SH [1].

6

Informal overview of the contents of this section: gas consumption; electricity consumption.

Chart depicting comparative natural gas usage. Notice this chart is treated as a figure and that the figure title appears below the chart.

Figure 1: Comparison of Annual Natural Gas Use by SH and EEH. [1, p. 8–9]

7

Figure 6-4 Pages from the body of an engineering report. Note the use of headings, bulleted lists, citations of borrowed information sources, and the chart.

SYMBOLS, NUMBERS, AND ABBREVIATIONS

Technical discussions ordinarily contain lots of symbols, numbers, and abbreviations. Remember that the rules for using numerals as opposed to words are different

Cross-reference to the table: Notice that explanation of the main trend in the table is provided.

Table title: Notice it is above the table and that a citation indicating the source is included.

First-level heading (using decimal numbering)

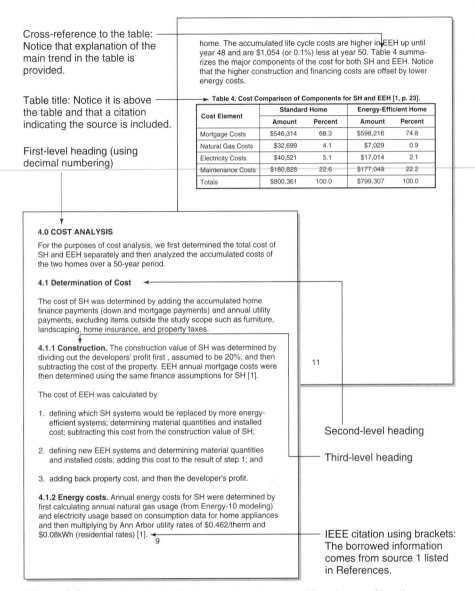

home. The accumulated life cycle costs are higher in EEH up until year 48 and are $1,054 (or 0.1%) less at year 50. Table 4 summarizes the major components of the cost for both SH and EEH. Notice that the higher construction and financing costs are offset by lower energy costs.

Table 4. Cost Comparison of Components for SH and EEH [1, p. 23].

Cost Element	Standard Home		Energy-Efficient Home	
	Amount	Percent	Amount	Percent
Mortgage Costs	$546,314	68.3	$598,216	74.8
Natural Gas Costs	$32,699	4.1	$7,029	0.9
Electricity Costs	$40,521	5.1	$17,014	2.1
Maintenance Costs	$180,828	22.6	$177,049	22.2
Totals	$800,361	100.0	$799,307	100.0

4.0 COST ANALYSIS

For the purposes of cost analysis, we first determined the total cost of SH and EEH separately and then analyzed the accumulated costs of the two homes over a 50-year period.

4.1 Determination of Cost

The cost of SH was determined by adding the accumulated home finance payments (down and mortgage payments) and annual utility payments, excluding items outside the study scope such as furniture, landscaping, home insurance, and property taxes.

4.1.1 Construction.
The construction value of SH was determined by dividing out the developers' profit first , assumed to be 20%, and then subtracting the cost of the property. EEH annual mortgage costs were then determined using the same finance assumptions for SH [1].

The cost of EEH was calculated by

1. defining which SH systems would be replaced by more energy-efficient systems; determining material quantities and installed cost; subtracting this cost from the construction value of SH;

2. defining new EEH systems and determining material quantities and installed costs; adding this cost to the result of step 1; and

3. adding back property cost, and then the developer's profit.

4.1.2 Energy costs.
Annual energy costs for SH were determined by first calculating annual natural gas usage (from Energy-10 modeling) and electricity usage based on consumption data for home appliances and then multiplying by Ann Arbor utility rates of $0.462/therm and $0.08kWh (residential rates) [1].

9

11

Second-level heading

Third-level heading

IEEE citation using brackets: The borrowed information comes from source 1 listed in References.

Figure 6-5 Pages from the body of an engineering report. Note the use of headings, tables, citations of borrowed information sources, and cross-references.

in the technical world. The old rule about writing out all numbers below ten does not always apply in engineering reports. (See Chapter 3 for discussion and guidelines.)

SOURCES OF BORROWED INFORMATION

To write your report, you may have to borrow facts and ideas from other engineers as well as from people in other professions. When you do, you must indicate the sources of your borrowed information, which is known as *documenting your sources*. See Chapter 11 for details.

GRAPHICS AND FIGURE TITLES

In an engineering report, you're likely to need drawings, diagrams, tables, and charts. These not only convey certain kinds of information more efficiently but also give your report an added look of professionalism and authority. If you've never put these kinds of graphics into a report, there are some relatively easy ways to do so—you don't need to be a professional graphic artist. See Chapter 7, "Constructing Tables and Graphics," for details.

CROSS-REFERENCES

You may need to point readers to closely related information within your report, or to other books and reports that have useful information. These are called *cross-references*. For example, they can point readers from the discussion of a mechanism to an illustration of it. They can point readers to an appendix where background on a topic is given (background that just does not fit in the text). They can also point readers to information outside your report—to articles, reports, and books that contain information related to yours. When you create cross-references, follow these guidelines (and see Figures 6-4 and 6-5 for examples):

- If you refer to another section of your report, put the heading or section title in quotation marks.
- If you refer to an article in a journal or encyclopedia, put quotation marks around the article title.
- If you refer to the title of a journal, book, or report, italicize that title.
- When you create cross-references, help readers understand why they should go to that information. Otherwise, they are likely to wonder. Indicate the topic of the cross-referenced information (don't assume the title indicates it fully), and suggest why readers might want to follow the cross-reference.
- Cite exact titles or supply page numbers if doing so helps readers. In a short report (say, one under 10 pages), citing page numbers is not necessary (although word processing software makes automating cross-references easy). If you supply the page number, then you can cite the subject matter of the section—not the exact title—in case you change the wording of headings.

CLARITY OF WRITING STYLE

As you rough-draft your report, don't get stymied over finding the exactly right words or avoiding grammar mistakes. In the rough-drafting stage, focus on the technical subject matter and don't get hung up on picky details that just slow you down. However, once you have a rough draft on paper or (more likely) in a computer file, reread it looking specifically for the common writing-style problems that make engineering writing, or any writing, hard to read.

Unnecessary Passive Voice. In the technical world, you must use the passive voice; but when it is misused, it leads to unclear, wordy writing. (See Chapter 3 for details on passive voice.)

Here's a typical example:

> In order to estimate company sales, industry estimates should first be looked at, because the sales of an individual company are often reflected by them.

This sentence could be rewritten as follows:

> To estimate company sales, look at industry estimates because individual company sales often reflect them.

Overreliance on the **Be** *Verb (Nominalization).* Heavy use of the *be* verb can make writing unclear and wordy as well:

> The User Name window is for entering the general information about the licensee (the customer). Contact your license administrator for defining the format.

Use active verbs instead. (See "Turning Verbs into Nouns" in Chapter 2.) In this example, how about "Use the User Name window to enter. . . "?

Unnecessary Expletives. Expletives use some form of *it is* or *there is*. They too can inflate writing, making it less direct and understandable.

> It is the results of studies of the central region of the M87 galaxy that have shown that there are stars near the center that move around as though there were some huge mass at the center attracting them.

Getting rid of the three expletives in the original produces this more readable version:

> Results of studies of the central region of the M87 galaxy show that stars near the center move around as though some huge mass at the center were attracting them.

Redundant Phrasing. For examples of wordy phrases and their concise counterparts, see the section on redundancy in Chapter 2. Here's a typical example:

> With reference to the fact that the company is deficient in manufacturing and production space, the contract may in all probability be awarded to some other enterprise.

The following revision leaves out the unnecessary words:

> The company may not be awarded the contract because it lacks production facilities.

Noun Stacks. Another problem, particularly in the technical world, involves jamming three or more nouns together into a phrase, which is called a *noun stack*.

> Cocombustion-chamber exit gas temperatures are approximately 2400°F.

In this example, why not say "The temperature of gas exiting the cocombustion chamber is about 2400°F"?

Weird Combinations of Subjects and Verbs. When you are struggling to express complex technical ideas, it's easy to combine subjects and verbs in strange ways, especially when lots of words come between them in the sentence. In the following example, it should be the *disappearance* that was *devastating*, not the *causes*:

> The causes of the disappearance of the early electric automobiles were devastating to the future of energy conservation in the United States.

These problems that create wordy, unclear writing are discussed in detail in Chapters 2 and 3, along with strategies for fixing them.

PARAGRAPH STRUCTURE

When you review your rough draft, look for ways to strengthen the organization and flow of your ideas. Do this kind of review at the level of whole paragraphs and whole groups of paragraphs:

- Strengthen transitions between major blocks of thought, such as between paragraphs or groups of paragraphs. (See Chapter 3 for more on transitions.)
- Add topic sentences (particularly the overview kind) to paragraphs where appropriate.
- Check the logic and sequence of paragraphs or groups of paragraphs. To do so, label each paragraph or paragraph group with one or two identifying words. This method enables you to get the global picture more easily.
- Break paragraphs that go on too long and challenge the reader's attention span.
- Consolidate clusters of short paragraphs that focus on essentially the same topic. Too many paragraph breaks can have a fragmented and distracting effect.
- Interject short overview paragraphs at the beginning of sections and subsections.

Using these strategies guides readers through your report, showing them what lies ahead, where they have come from previously in the report, and how everything fits together.

GRAMMAR, USAGE, AND PUNCTUATION

As mentioned earlier, you don't want to slow yourself down worrying about subjects and verbs, commas, apostrophes, and the like while writing. Worry about these details in the revising and editing stage. However, once you have a rough draft on paper or on disk, check for the various common mistakes such as those involving commas, apostrophes, spelling (particularly spelling of similar-sounding words), parallelism, agreement, and so on. See Chapter 3, "Eliminating Intermittent Noise in Writing," for details.

CONCLUSIONS

For most reports, you'll need to include a final section. When you plan the final section of an engineering report, think about the functions it can perform in relation to the rest of the report:

- *Conclude*. It draws logical conclusions from the discussion that has preceded; it makes inferences on the data that has been presented.

- *Summarize.* It reviews the key points, key facts, and so on from what has preceded. Summaries present nothing new—they leave readers with a perspective on what has been discussed, the perspective that the writer wants them to have.

- *Generalize.* It moves away from the specific topic of the report to a general discussion of implications, applications, and future developments—but only in general terms.

Your final section can do any combination of these, depending on your sense of what your report needs. The example conclusion in Figure 6-6 summarizes the key conclusion contained in the report, speculates about housing trends, and takes a brief look at recent developments.

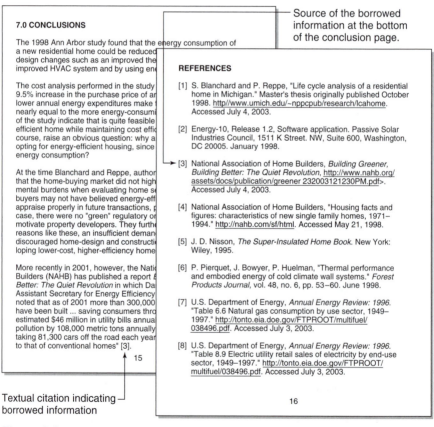

Figure 6-6 Conclusion and references page. Notice that the conclusion (which precedes the references pages) summarizes the chief finding of the report, speculates on that finding, and then glances at more recent developments. The references page uses the IEEE system of documenting borrowed information. Also see the bracketed citations in Figures 6-4 and 6-5. (See Chapter 11 for details on the IEEE system.)

The length of the conclusion can be anything from a 100-word paragraph to a 5- or 6-page section. For the typical 10- to 20-page report, the final section is one to two pages, but such ratios should never be applied rigidly without considering the report. Watch out for conclusions that get out of hand and become too long. Readers expect a sense of closure, a feeling that the report is ending. When the final section becomes too long, consider doing one of the following: Move the discussion back into the body of the report; shorten and generalize the discussion and keep it in the conclusion; or find some other way to end the report.

APPENDIXES

Appendixes are those extra sections following the conclusion. What do you put in appendixes? Anything that does not comfortably fit in the main part of the report but cannot be left out of the report altogether. The appendix is commonly used for large tables of data, big chunks of sample code, foldout maps, background that is too basic or too advanced for the body of the report, or large illustrations that just do not fit in the body of the report. Anything that you feel is too large for the main part of the report or that you think would be distracting and interrupt the flow of the report is a good candidate for an appendix.

DOCUMENTATION

Documentation is the system by which you indicate the sources of the information you borrow in order to write an engineering report. Many engineers use the system created by the Institute of Electrical and Electronics Engineers (IEEE), examples of which are shown in the figures throughout this chapter. Other engineering documentation systems vary only slightly from the IEEE one.

Documenting your information sources is all about establishing, maintaining, and protecting your credibility in the profession. You must cite ("document") borrowed information regardless of the shape or form in which you present it. Whether you directly quote it, paraphrase it, or summarize it—it's still borrowed information. Whether it comes from a book, an article, a diagram, a table, a web page, a product brochure, or an expert whom you interviewed in person—it's still borrowed information.

See Chapter 11 for details on how to cite the sources of your borrowed information using the IEEE documentation system.

EXERCISES

Look at some examples of technical reports to answer the following questions.

1. How does the format of these engineering reports compare with the format shown in this chapter or with that specified by the American National Standards Institute's *Scientific and Technical Reports: Organization, Preparation, and Production*?

2. What are the common audiences for the reports? Are they fellow engineers or non-specialists?

3. Typically, what purposes do the reports have? What functions do they perform for the engineering firm?

4. How were the graphics that are present in the reports created—by graphics specialists or by the engineers themselves?

5. How much are the reports a product of team writing—a group of engineers working on the project together?

6. How much library research is typically required to produce the reports? How much information for the reports comes from print and nonprint sources?

7. What process do engineering firms use in the production of reports? Do they use technical writers, graphics specialists, document designers, and editors, or is the production of reports mostly the responsibility of the engineers and clerical staff?

BIBLIOGRAPHY

American National Standards Institute. *Scientific and Technical Reports: Organization, Preparation, and Production*. ANSI Standard No. Z39.18-1987. New York: ANSI, 1987.

Beer, David. *Writing and Speaking in the Technology Professions: A Practical Guide,* 2nd ed. New York: IEEE Press, 2003.

Institute of Electrical and Electronics Engineers. *IEEE Information for Authors*. Piscataway, NY: IEEE Press, 1966.

7

CONSTRUCTING TABLES AND GRAPHICS

Too often writers overlook the importance of including graphics in their reports and papers. Correctly done, graphics (or visuals) not only are informative, but they also draw the readers' attention to information writers choose to highlight. They can carry much more information per space in a document than the same amount of text can. And if one definite trend is emerging in writing about high-tech subjects, it is an increasing reliance on visual communication.

Charles Sides, *How to Write and Present Technical Information*, 3rd ed. (Phoenix: Oryx Press, 1999), p. 48

When you write engineering documents, you're likely to need tables, illustrations, diagrams, charts, graphs, drawings, and schematics. Nontextual material like this helps present your information more effectively and gives a polished, professional look to your work. With the increasing power and ease of use of graphics software applications, you don't need to be a graphics professional to create or adapt graphics for your engineering documents.

If you're new to incorporating tables and graphics in engineering documents, consider the array of choices you have. Tables, graphs, and charts enable you to show data and, in particular, to show comparisons of data or changes in data.

- *Tables:* Rows and columns of numbers and words.
- *Graphs:* Represent data using lines that creep up and down from left to right, indicating changes in the data across time.
- *Charts:* Use bars, pie slices, or other inventive means to enable comparisons of data. The most common types are bar charts and pie charts.

The following are all illustrative graphics, intended to represent physical things:

- *Photographs:* Supply lots of detail—in some cases, too much. They are useful, for example, when you want to show a model of a new product.

- *Drawings:* Simplified illustrations of objects, people, and places. You see plenty of drawings used in instructions. They strip away extraneous detail and focus on the key objects and actions.

- *Diagrams:* Abstract illustrations of objects. They focus on infrastructural matters such as circuitry or piping and are often accompanied by measurements and symbols. Diagrams can also be used to illustrate nonphysical things such as concepts. An organizational chart of a company is a typical example. A flowchart of a production process is another.

TABLES

You've probably constructed tables using word processing applications such as WordPerfect or Word. This section provides some ideas for increasing your productivity with tables and for fine-tuning the design of tables. (See Figure 7-1 for table terminology.)

Table title (above table)

Table 4. Ozone levels for Houston and El Paso 1993

Row headings

Column headings (centered)

Month	Houston	El Paso
January	92	98
February	97	97
March	146	89
April	176	65
May	166	94
June	126	84
July	138	97
August	231	138
September	197	94
October	154	135
November	101	111
December	64	70

Right-aligned numeric data columns (but centered in the column as a group)

Measurement indicator (not repeated in every data cell)

Note: Measurements in parts per billion

Figure 7-1   Table terminology. You might prefer a table design with fewer grid lines, such as the table shown in Figure 7-4. Check your word processing software; it provides many different design options for tables.

CONVERT TEXT TO TABLES

In most word-processing software, you can convert a column of text to a table. Just make sure that you have a repeating set of elements: for example, a set of four repeating elements to create a four-column table (Figure 7-2).

USE TABLES FOR TWO-COLUMN LISTS

You have probably seen two-column lists and perhaps even created some by using tabs. Bad idea: When you add or delete words, the formatting falls apart. Instead, use a table in which you turn the grid lines off (Figure 7-3).

Energy Source
CO_2 (lb/kWh)
SO_2 (lb/kWh)
NO_x (lb/kWh)
Coal
2.12
0.0136
0.0079
Natural Gas
1.34
0.000007
0.0046
Oil
1.96
0.0123
0.0036

Table 2. Gas emissions per kilowatt-hour generated [21]

Energy Source	CO_2 (lb/kWh)	SO_2 (lb/kWh)	NO_x (lb/kWh)
Coal	2.12	0.0136	0.0079
Natural Gas	1.34	0.000007	0.0046
Oil	1.96	0.0123	0.0036

Figure 7-2   Converting text to tables. Notice that the text column is arranged in groups of four. (The table title is added afterward.)

cantilever beam	Projecting beam or member supported at one end.
current-factor	Rating system for current in transistors.
logic circuit	Circuits made up of transistors, diodes, and resist... AND,
polymers	Chem comp units

cantilever beam	Projecting beam or member supported at one end.
current-factor	Rating system for current in transistors.
logic circuit	Circuits made up of transistors, diodes, and resistors. The five common logic gates are AND, OR, NOT, NAND, NOR gates.
polymers	Chemical compound or mixture of compounds consisting of repeating structural units.

Figure 7-3   Two-column lists—an easier way. The version on the right is still the same table; its grid lines are turned off.

IMPORT SPREADSHEET DATA TO CREATE TABLES

Many of your tables may come from data in spreadsheet applications such as Lotus or Excel. There's no sense in retyping all that data—copy or import it instead. In most spreadsheet applications, copying is easy: Just select the cells you want, copy them, and then paste them into your document. In most applications, the pasted data cells will be formatted as a table; all you have to do is fine-tune the formatting (Figure 7-4).

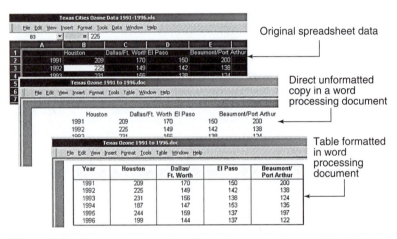

Figure 7-4 Using spreadsheets for tables. After pasting the spreadsheet data into a word processing document, you must format it, as shown here.

CONVERT PARAGRAPHS TO TABLES

Study your drafts for opportunities in which plain textual discussion in paragraph format can be reworked as tables. As you can see in Figure 7-5, discussions of two or more items in which the same categories of statement are made about those items are excellent opportunities for re-presentation as tables.

FORMAT TABLES

Whichever technique you use to create tables, keep these design considerations in mind:

In a comparison of Ford conventional vehicles and hybrid electric vehicles (HEV), the HEV proved to have a greater range (450–550 miles) than did the conventional vehicle (350 miles). And, as might be expected, these numbers were the same for gasoline range. In terms of fuel economy, the HEV was 30–50% better than the conventional vehicle. This, in turn, meant less frequent fill-ups for the HEV. Burning less gasoline causes the HEV to be 95% cleaner— far friendlier to the environment. And finally, this study found that the HEV performed more like a V-6 (more powerfully) than the conventional vehicle, whose performance was considered more like that of a 4-cylinder engine.

Table 1 shows the results of a comparison of conventional and hybrid electric vehicles done by Ford in 2002:

Table 1. Conventional-HEV Vehicle Comparisons

	Conventional	**Hybrid Electric**
Total Range	350 miles	450–550 miles
Gasoline Range	350 miles	450–550 miles
Fuel Economy	Base	30–50% over base
Re-fueling	Fill-up	Fill-up (less often)
Environmental Friendliness	Base	SULEV (95% cleaner than today's standard)
Performance	4-cylinder	Like a V-6

Source: Ford Motor Company. "Hybrid Vehicles," <www.ford.com/en/ourVehicles/ environmentalVehicles/hybridElectricVehicles/>. Accessed October 6, 2002.

Figure 7-5   Transforming text into a table. In the original version, data is buried in the textual discussion; in the revised version, it is taken out of paragraph format and presented as a table, making it more quickly scannable and breaking up the text.

- Include a heading at the top of each column to identify the contents of the column.
- If necessary, include a row heading in the farthest left column to identify the contents of the row.
- For textual material, left-align column headings and column contents.
- For numeric material, right-align column contents and center these column contents under the column heading.
- For any narrow stream of characters (numbers, letters, symbols), left-align column contents and center these column contents under the column heading.
- Put the measurement value in the column or row heading, not in each of the data cells.
- Put table titles *above* tables, not below. Use the word "Table," not "Figure." Notice that table titles can either be separate from the table or be the first row of the table, spanning all columns.

For other guidelines on tables, see the information presented by Jeffrey Donnell (Woodruff School of Mechanical Engineering, Georgia Tech) at http://fbox.vt.edu/ eng/mech/writing/handbook/visuals/donnell/tables/sld001.htm. Also see the sources listed at the end of this chapter.

CITE THE SOURCES OF TABLES

Whether you screen-capture a table from someone else or use only portions of some-one else's table, you still must document it—that is, indicate the source of that table. It's perfectly legal to copy a table verbatim from another source into your own engi-neering document, as long as you document its origins. Notice how the source of the table is indicated in Figure 7-5. However, you can also use the citation style of your documentation system—for example, the number of the source in brackets, as is done in Figure 7-2.

CHARTS AND GRAPHS

The terms *chart* and *graph* encompass the numerous ingenious ways of showing relationships between data—for example, line graphs, bar charts, pie charts, and three-dimensional variations such as pictographs. All of these types are visual repre-sentations of tables. They express a fundamental frustration with the dull old table— it is made up of row upon row and column upon column of numbers and words.

In tables, the significance of the data is not immediately evident without careful study. Charts and graphs, on the other hand, make that significance stand out. For example, if your department has reduced defects in the manufacturing process each year over the past five years, a line graph shows this point more vividly than a table. If those defects are primarily the result of faulty raw materials, then a pie chart depicts this fact much more strikingly than a table. See Figures 7-6 and 7-7 for illus-trations of how charts can present data more dramatically than tables.

How do you choose which type of graph or chart to use? Here are some ideas:

- *Line graphs* depict change in data occurring over time. Several lines enable readers to compare changes between different sets of data over time. Imagine a line graph showing total sales for Dell, Hewlett-Packard, IBM, and Apple over the past several decades. Figure 7-7 uses a line graph to show which city in Texas has the most ozone and how levels have fluctuated over the years.

- *Pie charts* depict the relative portion of a total amount made up by each member that contributes to that total. Pie charts give readers a dramatic sense of the percentages of each element making up a whole. Imagine a pie chart of total sales for Dell, Hewlett-Packard, IBM, and Apple in the year 2004. Who

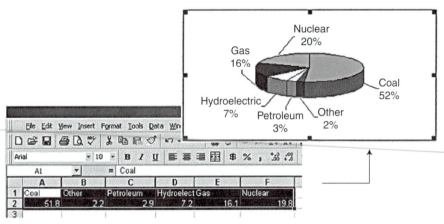

Figure 7-6 Pie charts from spreadsheets. This pie chart was created first by entering data into a spreadsheet application and then by choosing Chart from the Insert menu and making the appropriate selections. Although the chart is initially placed in the spreadsheet, you can copy it like any other object and paste it into a document.

would have the biggest slice? Figure 7-6 shows which energy source constitutes the biggest slice.

- *Bar charts* enable comparisons such as those shown in Figure 7-8. Bar charts can also, to a limited degree, indicate change over time. Imagine a bar chart showing sales for Dell, Hewlett-Packard, IBM, and Apple for 2004. However, time can be added: A set of four grouped bars for the sales of these companies

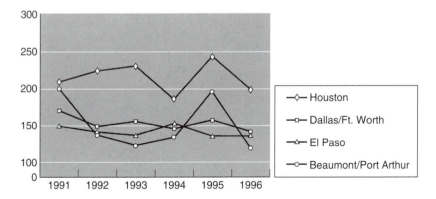

Figure 6. 1991–1996 ozone data for Texas cities (parts per million). [7]

Figure 7-7 Line graph using the same data as in Figure 7-4. Spreadsheet applications can also produce line graphs like this one. Notice that the title for this figure is located *below* the figure. Notice also that the source is indicated using the IEEE style of citation (see Chapter 11 for details).

could be created for each year (or decade). These sets can then be loaded into the same bar chart, enabling readers to see how these companies' sales compared in any given time period, how an individual company's sales changed over time, and how these companies' sales changed compared with each other over time.

For comparisons that more accurately demonstrate the performance of a dual processor system, VeriTest used the "SPEC rate" metrics, which recognize multiple processors. With SPECint_rate_base2000 and SPECfp_rate_base2000, the benchmark code is compiled and multiple copies are run concurrently, allowing both processors to work in parallel. SPEC rate tests determine the number of times a system can complete the benchmark per hour, also referred to as system throughput.

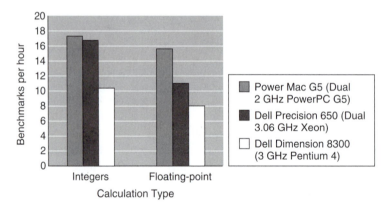

Figure 7. SPEC CPU2000: Dual Processor System Throughput. Integer calculations using SPECint_rate_base2000 and floating-point calculations using SPECfp_rate_base2000.

The results shown in Figure 7 demonstrate the benefits of the dual processor Power Mac G5. With full support for symmetric multiprocessing, dual independent 1 GHz frontside buses, and two floating-point units per processor, the dual 2 GHz Power Mac G5 completed the set of floating-point calculations 95 percent faster than the Pentium 4-based system and 42 percent faster than the dual Xeon-based workstation. Integer performance was also far superior to the Pentium 4-based system and 3 percent faster than the dual Xeon-based system.[5]

[5] Based on SPEC CPU2000 benchmark results against 3 GHz Pentium 4-based Dell Dimension 8300 and dual 3.06 GHz Xeon-based Dell Precision Workstation 650m, performed by VeriTest, June 2003.

Figure 7-8 Clustered bar chart. Notice how this example uses multiple bars to show multiple comparisons of processor speed. Notice too that the nearby text explains the significance of the bars. *Source:* a192.g.akamai.net/7/192/51/ebb34a6c95daa5/www.apple.com/powermac/pdf/PowerMacG5_Perf_WP_062303.pdf.

- *Tables*, on the other hand, enable the number-crunchers and the bean-counters to do their jobs. Charts and graphs are generally useless to people entering numbers into electronic spreadsheets or databases.

Once again, many other issues involving graph and chart construction are covered by Jeffrey Donnell (Woodruff School of Mechanical Engineering, Georgia Tech) at http://fbox.vt.edu/eng/mech/writing/handbook/visuals/donnell/graphs/sld001.htm as well as the sources listed at the end of this chapter.

ILLUSTRATIONS

As mentioned earlier, the term *illustration* refers to all manner of pictorial graphics—photographs, drawings, diagrams, and schematics. Included here are also conceptual diagrams such as flowcharts, even though they represent physical reality only in the most symbolic way.

If you must illustrate something in your engineering documents, consider carefully whether you need a photograph or some type of diagram. Photographs provide the greatest amount of visual detail; however, that may be too much detail, for example, in instructions for upgrading the microprocessor in your computer. Diagrams omit unnecessary detail and enable readers to focus on the essentials (Figure 7-9). Diagrams can range from the closely pictorial (see the microprocessor diagram in Figure 7-10) to the highly abstract (see the conceptual graphic in Figure 7-11). Choosing the type of illustration depends on which works best for the reader and the purpose of your document.

But how do you create or acquire illustrations in the first place? Here are some starting points:

- *Internet.* Search the Web for illustrations. When you find one you like, download it or copy it by means of a screen capture. Don't forget to copy the URL, the title of the web page, the author's name, and the date the illustration was created, if available. Also, make a note of the date you accessed that page. (See Chapter 11 for details on citing sources of graphics.)
- *Hardcopy scans.* Scanning equipment has improved immensely; you can scan an image right from a print document. All the rules of citing the source from which you borrowed the image still apply. (Again, see Chapter 11.)
- *Professional clipart.* You may also be able to purchase a CD or DVD loaded with engineering and scientific clipart. Among these generic drawings you should be able to find the graphics you need.
- *Graphics professional.* If your budget allows, you can outsource the work to a graphics professional who creates technical drawings for a living.

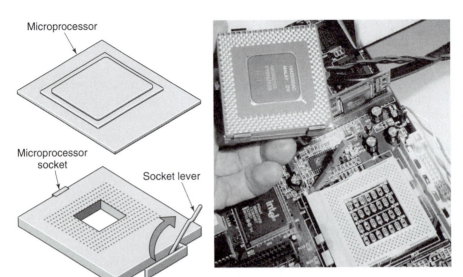

Figure 7-9   Diagrams and photographs. Getting a photograph with good detail like the one here is difficult. More often, a simple line drawing like the image on the left is clearer and more understandable for readers. (Photograph reprinted with permission from ThePC.info, "How do I upgrade my microprocessor?" www.thepc.info/CPU_upgrade.html. Diagram reprinted with permission from Dell Computer Corporation, "Dell Precision Work-Station 530 User's Guide," support.jp.dell.com/docs/systems/ws530/en/ug/html/2prsr.htm.)

- *"Low-tech" graphics production.* If none of the methods discussed previously works, and if you consider yourself a horrible artist, there's still hope:

  1. Find the graphic that you want in a book, report, or journal, for example. Avoid graphics in low-quality print media such as newspapers; they won't photocopy well. Photocopy the graphic, and enlarge or reduce it as needed.

  2. Trim the copy, cutting out the figure title, and legends, but not necessarily the labels (usually they will work just fine as your own labels). As necessary, add the titles and legends yourself—either in your word processing software or your graphics software—to make the graphic functional in your engineering document.

  3. In the text of your engineering document, decide where to place the graphic—ideally, just after the point at which it is referenced. Leave enough space above and below the graphic so that it won't appear squeezed in. Make sure it doesn't spill outside your regular right and left margins.

  4. In your file (or typed page), type the figure title. Because you've photocopied the graphic, you must cite the source, just as with any information you borrow. (See Chapter 11 for details.)

Additionally, IBM and Apple added processing innovations to the PowerPC G5 that optimize the flow of data and instructions, enabling the PowerPC G5 to process 71% more instructions at a time than the 32-bit Pentium 4:

1. *L2 Cache.* 512 K of L2 cache (see Figure 4) provide the execution core with ultrafast 64-MBps access to data and instructions.

2. *L1 Cache.* Instructions are prefetched from the L2 cache into a direct-mapped 64 K L1 cache at 64 GBps. In addition, 32 K of L1 data cache can prefetch up to eight active data streams simultaneously.

3. *Fetch and Decode.* As they are accessed from the L1 cache, up to eight instructions per clock cycle are fetched, decoded and divided into smaller, easier-to-schedule operations. This preparation increases processing speed as instructions are dispatched into the execution core and data is loaded into registers behind the functional units.

4. *Dispatch.* Before instructions are dispatched into the functional units, they are arranged into groups of up to five. Within the core alone, the PowerPC G5 can track up to 20 groups at a time, or 100 individual instructions. This scheme enables the PowerPC G5 to manage an unusually large number of instructions "in flight": 20 instructions in each of the five...

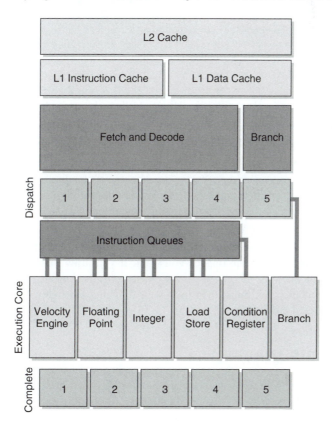

Figure 7-10 Architectural diagram of a microprocessor. This diagram is not only highly abstract but also "conceptual" in that the physical PowerPC G5 does not resemble this diagram at all—however, in terms of its hierarchy of functions and components, it does. *Source:* www.apple.com/g5/executioncore.html.

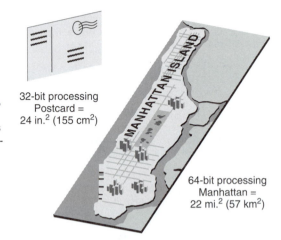

32-bit registers give a processor a range of 2^{32}, or 4.3 billion—which means it can express integers from 0 to 4.3 billion. With 64-bit registers, the dynamic range catapults to 2^{64}, or 18 billion billion—4.3 billion times larger than the range of a 32-bit processor (the difference between the size of a postcard and the size of Manhattan island).

32-bit processing
Postcard =
24 in.^{2} (155 cm^2)

64-bit processing
Manhattan =
22 mi.^{2} (57 km^2)

MANHATTAN ISLAND

Figure 7-11   Conceptual graphic designed by Apple Computer, Inc. to provide a sense of the advance that 64-bit processors represent over 32-bit processors. *Source:* a1264.g.akamai. net/7/1264/51/d297fb89c825b9/www.apple.com/g5/pdf/G5_Processor_WP_062303.pdf.

5. When you've printed or typed your engineering document, carefully place your graphic in the space you've left open for it, and then tape or glue it, making sure that the seams and the tape won't show in the photocopy.

6. Finally, get a good-quality photocopy not only of this page but of all the pages in your engineering document. Never submit an engineering document with things taped, stapled, or clipped onto the pages. (And don't draw or color on your final pages; if that's what you want, use a color photocopier.)

It may not be so out of the question for you to create some of your graphics yourself. Consider tracing the images you want. If you draw freehand, use a soft pencil and light marking to get the drawing just right, then ink it in with a black marker. Erase your pencillings, then treat your drawing just like the photocopied graphics discussed previously.

ELECTRONIC IMAGES

If you work directly with electronic images, be sure you know how to do these tasks in a graphics application such as Adobe Photoshop, Adobe Illustrator, CorelDRAW, Jasc Paint Shop Pro, or other similar application:

*   *Crop.* Know how to trim away unwanted material from the graphic.

*   *Size.* Know how to enlarge or reduce a graphic, and understand the distortion that occurs when you do so.

- *Label.* Know how to add textual labels and arrows to a graphic.
- *Clean up.* Know how to erase minor details from a graphic and how to sharpen, add contrast, darken, or lighten a graphic.

GRAPHICS AND TABLES: GUIDELINES

When you incorporate graphics and tables into an engineering document, pay attention to their standard components, their placement, and cross-references to them. The following list summarizes guidelines stated throughout this chapter.

- *Add figure and table titles.* Include descriptive figure titles below illustrations, diagrams, charts, and graphs. Include descriptive table titles above tables. Titles identify the content of figures and tables at a glance for readers who are scanning.
- *Add labels.* In illustrations, add words that identify the parts of the thing being illustrated, and a pointer from each label to the part being illustrated. In charts and graphs, add labels to the axes to identify the units of measurement and other details.
- *Indicate sources of borrowed graphics or tables.* See Chapter 11 on methods of documenting your borrowed tables and graphics. It's easy to grab material from the Web, but remember to copy the URL, page title, any information on author and date updated, and the date *you* accessed the page.
- *Place graphics and tables at the point of first reference.* Position graphics and tables just after the first point in your text where they are referenced. If they don't fit on the same page, place them at the top of the next. Each graphic or table should appear as soon as possible after you first mention it.
- *Align and position graphics carefully.* Maintain adequate spacing between graphics and text; make sure that graphics are visually nicely balanced on your pages (Figure 7-12a). For example, if you create a graphic less than a half-page in size, you can have your text flow around it (Figure 7-12b). Don't cramp things, however. Make sure you leave plenty of white space between your text and graphic and that your graphics fit within your regular margins.
- *Intersperse graphics and tables with text.* Insert graphics and tables into pages with text rather than appending them at the end of the document. For readers, it's pleasing to have text broken up with graphics and tables.
- *Include a legend.* If your graphs or charts use different symbols, colors, shadings, or patterns to indicate different elements, include a legend. See Figures 7-7 and 7-8 for examples of legends.

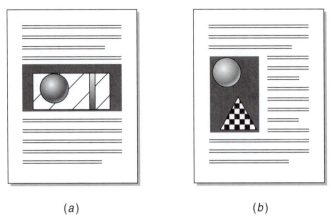

(a)   (b)

Figure 7-12   (*a*) Example of effective centering of a graphic on the page. (*b*) Letting text flow around a graphic can give your page a professional look.

- *Provide cross-references to your graphics and tables.* Don't just pitch graphics and tables into engineering documents without referring to them and explaining key points. Otherwise, readers may have a nice picture or a pile of statistics, but no sense of the purpose or meaning. Use phrasing like the following:

As can be seen in Figure 5, the thermophysical properties . . .

The arrangement of the MOF network (Fig. 8.2) is structured so that . . .

Averages for the fabric cutting speeds are shown in Table 4 on the next page.

EXERCISES

1. Find a relatively simple table and reconstruct it in your own software application using the techniques and guidelines discussed in this chapter.
2. Find a relatively simple table in which the data can be converted to a line graph. Create the line graph using the techniques and guidelines discussed in this chapter.
3. Find a relatively simple table in which the data can be converted to a bar chart. Create the bar chart using the techniques and guidelines discussed in this chapter.
4. Find a relatively simple table in which the data can be converted to a pie chart. Create the pie chart using the techniques and guidelines discussed in this chapter.
5. Find text with illustrations (photographs or diagrams) on the Web, and reconstruct that page, including the illustrations, in your own software application using the techniques and guidelines discussed in this chapter.

BIBLIOGRAPHY

Donnell, Jeffrey. "Presentation Visuals on Illustrating Technical Documents." http://fbox. vt.edu/eng/mech/writing/other/illustration.html. Accessed July 2, 2003.

Horton, William. "The Almost Universal Language: Graphics for International Documents." *Technical Communication* 40 (1993), 682–693.

Horton, William. "Overcoming Chromophobia: A Guide to the Confident and Appropriate Use of Color." *IEEE Transactions on Professional Communication* 34 (1991), 160–171.

Lefferts, R. *How to Prepare Charts and Graphs for Effective Reports.* New York: Barnes and Noble, 1982.

Lichty, T. *Design Principles for Desktop Publishers.* Glenview: Scott, Foresman, 1989.

Lo, Jack, and David Pressman. *How to Make Patent Drawings Yourself: Prepare Formal Drawings Required by the U.S. Patent Office.* Berkeley, CA: Nolo Press, 2002.

MacGregor, A. J. *Graphics Simplified: How to Plan and Prepare Effective Charts, Graphs, Illustrations, and Other Visual Aids.* Toronto: University of Toronto Press, 1979.

Robertson, B. *How to Draw Charts and Diagrams.* Cincinnati: North Light Books, 1988.

Tufte, Edward R. *Envisioning Information.* Cheshire, CT: Graphics Press, 1990.

Tufte, Edward R. *The Visual Display of Quantitative Information.* Cheshire, CT: Graphics Press, 1992.

Tufte, Edward R. *Visual and Statistical Thinking: Displays of Evidence for Decision Making.* Cheshire, CT: Graphics Press, 1997.

White, Jan. *Visual Design for the Electronic Age.* New York: Watson-Gupthill, 1988.

8

Accessing Engineering Information

Most staff members realize that the [company] library exists and maintains a collection of reference works, but many don't know that modern computer technology has made the library a formidable reference identification and retrieval system.

Richard Manley, et al., "Some Guidance on Preparing Technical Articles for Publication," *IEEE Transactions on Professional Communication (March 1989), p. 9*

I was gratified to be able to answer promptly, and I did. I said I didn't know.

Mark Twain

You probably don't need to be reminded that scientific information is growing at breakneck speed—according to some estimates doubling every two or three years—while the electronic pathways to this knowledge are also rapidly expanding. The information explosion is now a constant state of affairs, and certainly a way of life for engineers. Moreover, boundary lines between science, engineering, and society are increasingly becoming blurred. For example, civil engineers, telecommunications engineers, and geologists may combine forces to build a very large antenna array near a city—and at the same time interact with biologists and the general public on such matters as the economic and environmental impact of their work.

Even if you work in a highly specialized field like these engineers, you may need to access information from fields other than your own. To support you in that effort, this chapter explores engineering information resources available for your reference and research.[1]

[1]Many thanks to Susan Ardis, head librarian, engineering library, University of Texas at Austin, for her work on the first edition of this chapter and to Teresa Ashley, MLS, Austin Community College, for her work on the second edition of this chapter.

BASIC SEARCH STRATEGIES

Before setting out for the library or opening your favorite Web search engine, know some strategies for planning your search and for getting the most out of your search.

PREPARING FOR THE SEARCH

Although books and journals are still important sources of information (and are usually what we associate with the traditional library), they are no longer the only sources we use. The 21st-century library is a hybrid of print and electronic resources. Since 1995, material has become increasingly and rapidly available on the World Wide Web. The Internet has become the vehicle for accessing libraries' subscription databases and indexes; electronic periodical indexes increasingly provide the full text of the magazine and journal articles that they index. Many libraries have growing collections of electronic books (e-books), and most publishers of print material offer their titles in both formats.

Even so, few engineers have the leisure to browse in a library or on the Web until they stumble on the right article, book, report, or website. When you need information, you should first spend time focusing on what it is you need and where it might be. Systematically ask yourself these questions:

What is my purpose?
- Write an internal report
- Work on a design problem
- Conduct research
- Select equipment or products

What kind of information do I need?
- Practical
- Theoretical
- Economic or public policy
- Proprietary
- Product information

What exactly do I need?
- Raw data
- Overview of the subject
- Historical information (for example, for product liability)
- Up-to-date, state-of-the-art information

- Competitive intelligence (what is our competition up to?)
- Intellectual property information
- Patents
- Trademark

What is my time frame?
- Hours
- Days
- Weeks
- Months ⸳

What information resources do I have access to?
- Nearby experts
- Publications that colleagues and I have stacked away
- Company library
- Electronic access
- Technical, college, university, or public library
- Technical book store

Am I willing to pay for the information?
- Buying relevant books
- Hiring a professional searcher to find what I need
- Paying for a full-text electronic search

Your answers to these questions determine where you will look for information. Remember: Practically any information is available if you have enough time and money. If you don't have such luxuries, however, being as specific as possible from the beginning of your search will help you reach your information goals.

FOLLOWING THE TRAIL

When you need background or history on a subject, start down the information trail with the most readily available tools first. These are usually technical encyclopedias, handbooks, books, and periodicals. Be aware of the publication dates of such sources: You are not likely to find much about laser surgery in anything printed before 1950; however, you would find plenty on helical springs and internal combustion engines.

To find specific rather than general information, be as precise as possible. Look for exactly what you want first; you can always become more general again later if necessary. To do that, figure out the hierarchy of your topic: what is more specific and what is more general. This technique works with all kinds of tools, including encyclopedias, book indexes, periodical indexes, and electronic access tools. If your topic is photovoltaic cells, for example, that hierarchy could be any of the subdivisions shown in Figure 8-1, depending on your focus.

If you are in a library and stray off the trail, don't hesitate to ask for help. Many libraries have staff who are experts in carrying out an information search and who are willing to assist you when courteously approached. Engineering librarians often suggest that you apply the 20-minute rule: If after looking for information for 20 minutes, you find nothing relevant, ask for reference help (in most libraries, this means a trained librarian, not a clerk).

Another part of staying on the information trail is to become proficient with the search engines on the Internet. An amazing amount of engineering data is now available from these sources; more information becomes available every day. (See "Internet Search Tools" later in this chapter for some starting points.)

SOURCES OF ENGINEERING INFORMATION

When you search for information for an engineering project, you are likely to use an array of information resources, including books, reference books, journals (as well as the indexes and abstracts associated with them), technical reports, patent literature, product literature, and specifications.

GENERAL BOOKS

In the United States alone, more than 50,000 book titles are now published annually, compared with 20,000 in 1960. Although all the information contained in these hard-

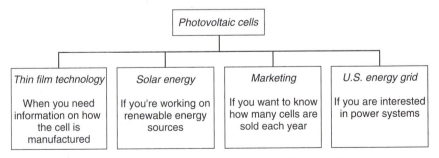

Figure 8-1 Hierarchies and subdivisions of information. In your information search, start as low in the hierarchy as you can (in this diagram, thin film technology, for example)—not high (photovoltaic cells).

copy publications could be made available electronically, this is not likely to happen for a good many years. Expect to find plenty of worthwhile information available only in hard copy only in library stacks and periodical rooms.

When to Use Books. When you are researching a topic, books can provide excellent background. A quick look through a book's table of contents and index will give you a good idea of whether it's likely to have what you are looking for.

Obviously, the most recently published books are going to give you the best picture of a current area of technology, but some older books may provide excellent background to a field. For example, a book or encyclopedia published in the 1960s on radio wave propagation and the ionosphere might still contain some useful background. For many current research topics, however, books tend to be too general. To obtain more specific information on technological advancements, you must go to journal articles, technical reports, or other sources (described later in this chapter).

How to Find Books. Library catalogs are online, making them accessible wherever there is a connection to the Internet. This means that if you cannot find a specific book or the right book in your library, you can check other libraries on the Internet. Online library catalogs offer you powerful search tools. Searching electronically by keyword, for instance, allows you to access several fields of the library catalog record at once: title, author, subject, and any notes that are available. (See Table 8-1 for a list of Internet-accessible libraries.)

How to Obtain Books. What if you find something in a library 400 miles away? There are two possibilities. Ask your librarian about an interlibrary loan. (If waiting for the book to arrive by mail is not an option, at least you know what is available on your topic.) However, you may find that many of the titles in your search results are electronic and can be read online. These e-books are included in library catalogs and can be checked out or browsed online. NetLibrary (www.NetLibrary.com) is probably the largest supplier of e-books, but many other publishers make content available online as well.

REFERENCE BOOKS

In addition to books located in the stacks, most engineering libraries (and technical information centers) have a reference section where you can find dictionaries, encyclopedias, handbooks, reference manuals, guides, and similar materials. These volumes normally cannot be checked out of the building and are used for quick, on-the-spot look-up of factual information. The best way to find useful reference tools is to use onsite or online access to library catalogs. If you don't know the URL (Web address) of a library's catalog, just type in the name of the institution in quotes and "library catalog" in the search field of a Web search engine, such as Google (www.google.com).

Table 8-1   Finding Internet-Accessible, Engineering-Related Libraries

American Society for Engineering Education, www.englib.cornell.edu/eld/libraries.php	The Engineering Libraries Division maintains a list of science, technology, and engineering libraries (mainly located in North America, but in a few other countries as well).
Grainger Engineering Library, gateway.library.uiuc.edu/grainger /resrc/englib/DEFAULT.asp	Maintained at the University of Illinois at Urbana-Champaign, this resource provides a directory of U.S. engineering libraries that can be browsed by state.
LibDex, www.libdex.com/	List of libraries (not necessarily engineering libraries) that allows keyword search and browsing by country.
The WWW Library Directory, www.webpan.com/msauers/libdir/	Lists over 8000 libraries worldwide.
Libweb, sunsite.berkeley.edu/Libweb/	Maintained at the University of California, Berkeley, this resource enables you to search libraries in over 115 countries using keyword searches or hyperlinks.
Library of Congress Catalog, catalog.loc.gov/	All books everywhere.

When to Use Reference Books.   As you can see from the examples in Table 8-2, reference books include a broad range of dictionaries, encyclopedias, and handbooks. In other words, they constitute a starting point or foundation for what is known in a field. Although new research may make some information in reference books obsolete, reference books on the whole represent standard knowledge and practice in their fields.

How to Find Reference Books.   Although the search interface varies from library to library, most have common features. You can always search by title or author, as well as by keyword. Keyword searching is best when you aren't sure what subject the book might be cataloged under and when you don't know the author or title.

- Any of the reference titles listed in Table 8-2 can be found by typing in a couple of words of the title (for example, mechanical engineer* handbook), truncating any words that could have variant endings (in other words, engineer* could retrieve *engineer*, *engineers*, or *engineering*) and eliminating any prepositions or articles (*of, for, the, a, an*).
- If you want to find an encyclopedia, type encyclopedia chemical or encyclopedia chemistry in the library catalog search field. This is the simplest and most efficient way to see what encyclopedias are available and what nearby libraries have available.

Table 8-2   Examples of Engineering Reference Books

Kirk-Othmer's *Encyclopedia of Chemical Technology*, 6th ed. 1992. 25 volumes. Covers all areas of technology—not just chemical. At the end of each article are useful references to patents, conference proceedings, and journal articles.

McGraw-Hill Encyclopedia of Science and Technology, 9th ed. 2002. 20 volumes. Contains almost 8000 well-written and well-illustrated articles on science, engineering, and other technical subjects. Check here first for general background.

Van Nostrand's Scientific Encyclopedia, 9th ed. 2002. 2 volumes. Concentrates on the basic and applied sciences, with over 17,000 articles. Also functions as a technical dictionary. Available online as AccessScience.

McGraw-Hill Dictionary of Scientific and Technical Terms, 6th ed. 2002. Provides more than 125,000 definitions of terms and includes some 3000 illustrations.

Encyclopedia of Energy Technology and the Environment, 1995. Part of the Wiley Encyclopedia Series in Environmental Science. Four volumes of articles on energy-related topics relating to technology and its impact on the environment.

Handbook of Industrial Engineering, 3rd ed. 2001. Almost 2900 pages of detailed information on such topics as performance measurement, quality control, engineering economy, and manufacturing engineering.

Marks' Standard Handbook for Mechanical Engineers, 8th ed. 1978. Continues the *Standard Handbook for Mechanical Engineers*. The 10th edition is available online by subscription at www.knovel.com/knovel2/Toc.jsp?BookID=346 and may be in some engineering library collections.

Perry's Chemical Engineers' Handbook, 7th ed. 1997. Includes material from general mathematics, tables, and specialized treatment of topics such as psychometry, process machinery, and distillation. A standard for petroleum and chemical engineers.

CRC Handbook of Mechanical Engineering. 1998. Like *Marks' Standard Handbook for Mechanical Engineers*, this handbook contains useful articles, tables, and data on all aspects of mechanical engineering and other subjects of use to mechanical engineers.

Standard Handbook for Civil Engineers, 5th ed. 2003. Covers construction, structural theory and design, materials, and management for the various fields of civil engineering, including environmental concerns.

Standard Handbook for Electrical Engineers, 14th ed. 1999. Substantial coverage of all aspects of electrical engineering, with numerous tables, charts, and graphs.

JOURNALS

You probably already subscribe to one or two professional journals and may have access to others through a local library. Over 10,000 hardcopy and hundreds of electronic scientific and technical journals are published every year, and both numbers are growing. These journals are essential for any engineer who wants to keep up with the latest developments in a given field.

When to Use Journal Articles. The information in journals (unlike books) consists of highly technical short papers and reports on the most current research and thinking in an area of specialization. Few libraries can subscribe to more than a fraction of the journals published, and most libraries either have a cross section of only the most respected journals or concentrate on journals for only a certain field. These limitations can be overcome by interlibrary loans, of course, if the journal isn't available electronically.

To become familiar with all the journals published in your field, consult *Ulrich's International Periodicals Directory*, which annually lists journal titles in some 200 categories, including engineering, which itself is further subdivided by fields such as civil, electrical, mechanical, and petroleum. Most public and college or university libraries own *Ulrich's* or have access to it by subscription online at www.ulrichsweb.com.

How to Find Journal Articles. The traditional way to find articles is to use a periodical index. But there are also some electronic options for journal research, such as CARL and WORLDCAT. Both are available over the Internet, and many libraries provide access through their subscriptions. CARL is particularly powerful because it consists of a searchable list of current journals (1986 to the present) and the tables of contents of those journals, and because it enables you to order copies of articles for a fee. WORLDCAT contains the titles of all books and journals owned by participating libraries.

How to Obtain Journal Articles. Not all scientific and engineering articles can be retrieved through traditional bibliographic sources such as periodical indexes. Within the past few years, there has been considerable growth in online collections of preprint articles, particularly in the sciences. Preprints are also referred to as *e-prints*. Though preprint articles have not yet been published, they may have been submitted, reviewed, and accepted for publication or made available online prior to presentation at conferences.

Engineers and scientists are interested in preprints because research findings can be made available sooner than through the traditional peer-review process and because articles can be circulated for comment prior to journal publication or conference presentation. Because wide readership is wanted and encouraged, these articles are often available free of charge to anyone who can access them. Because these articles are not published in journals, they cannot be found by searching periodical indexes. Preprint networks provide access to them. Be aware of two preprint servers:

- The arXiv e-print archive (for articles in physics, mathematics, nonlinear science, and computer science) from Cornell University, accessible at arxiv.org/.
- The E-print Network (for engineering documents) operated by the Department of Energy (DOE) Office of Scientific and Technical Information (OSTI), accessible at www.osti.gov/eprints/.

Indexes and Abstracts

If you intend to use research articles for an engineering document, you must be aware of two essential tools for finding and selecting those articles: indexes and abstracts.

Indexes. Imagine that you wanted to find all the research articles written on fuel cells before President George W. Bush announced his intentions for the Department of Energy to refocus its work on that technology in 2002. How would you find all those articles? Spend weeks scanning the table of contents of likely journals? No, instead you'd use a *periodical index* to find articles on fuel cells published in a wide array of magazines and journals.

An index lists articles grouped by subjects from selected periodicals. For each article, you'll find the article title, author(s), periodical title, volume, issue, date, and page count. Some indexes include abstracts (summaries) of each article. Indexes are available for broad categories or fields, as shown in Table 8-3.

You can find indexes in libraries in both print and electronic form. Most electronic indexes cover articles published after 1984 or so, although many are adding earlier publications retrospectively. If you need an article published before an electronic index, use the print volumes, which can go back as far as the early 20th century.

Abstracts. As you know, periodical indexes enable you to find articles on a specific topic. But what if these indexes enable you to find too many articles—far more than you can read in time to complete your project? You read the *abstract* of each article to decide whether to use that article or not. Abstracts appear not only at the top of the

Table 8-3   Examples of Engineering-Related Indexes and Abstracts

Title of Paper Index	Electronic Equivalent
Applied Science and Technology Index	ASTI or ASTA
Business Information	ABI/Inform
Chemical Abstracts	CA
Computer and Control Abstracts	INSPEC
Electrical and Electronics Abstracts	INSPEC
Engineering Index	Compendex
International Aerospace Abstracts	Aerospace Database
Metals Abstracts	Metadex
Nuclear Science Abstracts (1946 to 1976)	Not available
Pollution Abstracts	Same name

When an index also contains the abstracts of the articles indexed, it is called "abstracts." These are just a few of the indexes and abstracts available in the field of engineering.

articles themselves but also separately in indexes. Most libraries provide a list of their electronic databases and online periodical indexes on their home page. For example, see the listing provided by Stanford at www-sul.stanford.edu/catdb/sci.html. Look particularly at the indexes and abstracts marked "Open to all." For example, select Astrophysics Data System Abstract Service and then Astronomy and Astrophysics. Figure 8-2 shows part of the search interface; Figure 8-3 shows what one of the index entries looks like, along with its abstract.

Unfortunately, few electronic indexing and abstracting services are "open to all" as is the Astrophysics Data System Abstract Service. You cannot access these resources unless you are affiliated with an institution that pays for (subscribes to) them, such as an academic or corporate library. When you use these resources away from the library, you supply a user ID and password to gain access.

TECHNICAL REPORTS

Hundreds of thousands of technical reports are written each year in the United States alone; many are available on electronic media. A technical report may be similar to a paper presented at a conference or to a journal article, but may be a lot longer. Technical reports are usually written by specialists for other specialists and report on the results of research and development.

Reports sponsored by a government grant or contract are the easiest to find, whereas proprietary and classified reports are not generally available. Because so many reports constantly spring up in the forest of technical information, you must use indexes and abstracts to narrow your search for reports in a certain field or for a specific report. This is one area where paper indexes are being replaced by electronic indexes such as NTIS and NTRS, described in Table 8-4. (See Figure 8-4 for an illustration of a typical NTIS record.)

Enter Abstract Words/Keywords   Require text for selection

(Combine with: ● OR   ○ AND   ○ simple logic   ○ boolean logic)

potatoes spinach wheat tadpoles fish mars space

Return 100   items starting with number 1

[Send Query]   [Return Query Form]   [Store Default Form]   [Clear]

Figure 8-2   Search terms entered at Astrophysics Data System Abstract Service, Astronomy and Astrophysics subcategory (www-sul.stanford.edu/catdb/sci.html). This is a search for research articles on the effects of extraterrestrial existence on animal and plant life.

Title:	Effects of modified atmosphere on crop productivity and mineral content
Authors:	Chagvardieff, P.; Dimon, B.; Souleimanov, A.; Massimino, D.; Le Bras, S.; Péan, M.; Louche-Teissandier, D.
Affiliation:	CEA, Direction des Sciences du Vivant, Départment d'Ecophysiologie Végétale et de Microbiologie, Centre de Cadarache, F-13108 Saint-Paul-Lez-Durance cédex, FRANCE
Journal:	Advances in Space Research, Volume 20, Issue 10, p. 1971-1974. (AdSpR Homepage)
Publication Date:	00/1997
Origin:	ELSEVIER
Abstract Copyright:	(c) 1997 Elsevier Science B.V. All rights reserved.
Bibliographic Code:	1997AdSpR..20.1971C

Abstract

Wheat, potato, pea and tomato crops were cultivated from seeding to harvest in a controlled and confined growth chamber at elevated CO_2 concentration (3700 muL.L^-1) to examine the effects on biomass production and edible part yields. Different responses to high CO_2 were recorded, ranging from a decline in productivity for wheat, to slight stimulation for potatoes, moderate increase for tomatoes, and very large enhancement for pea. Mineral content in wheat and pea seeds was not greatly modified by the elevated CO_2. Short-term experiments (17 d) were conducted on potato at high (3700 muL.L^-1) and very high (20,000 muL.L^-1) CO_2 concentration and/or low O_2 partial pressure (~ 20,600 muL.L^-1 or 2 kPa). Low O_2 was more effective than high CO_2 in total biomass accumulation, but development was affected: Low O_2 inhibited tuberization, while high CO_2 significantly increased production of tubers.

Figure 8-3 Abstract—example. This abstract is typical of what you see in electronic indexing and abstracting services. You get both the index entry with bibliographic detail to enable you to find the complete article, plus the abstract, which provides a summary of the research purpose and outcomes.

PATENTS

Patent documents are a rich source of technical and scientific information. They describe in detail the designs, materials, machines, and processes associated with inventions. In return, the government grants the inventor a right of ownership that limits others from making, using, or selling the patented item in the United States for 20 years. In 1790, the year it was created, the U.S. Patent Office granted 3 patents; these days, the number of U.S. patents granted annually is close to 150,000. As of 1995, some 5.5 million patents had been issued. As Table 8-5 shows, it's amazing what gets patented.

Table 8-4 Finding Engineering Reports

National Technical Information Service (NTIS), www.ntis.gov/	Major source for information on nonproprietary and unclassified reports sponsored by government agencies and contractors. NTIS lists the subject of each report, its individual and corporate author, and the contract and report number. You can search technical reports on government-sponsored research from organizations such as NASA, DOE, and EPA. You can read abstracts for the reports online; the reports can be purchased online.
NTRS (NASA Technical Reports Server), ntrs.nasa.gov/	Collects, archives, and makes available NASA's scientific and technical information, including research reports, journal articles, conference and meeting papers, technical videos, mission-related operational documents, and preliminary data. Available via the NASA Technical Report Server (NTRS) to provide students, educators, and the public access to NASA's technical literature.
IEEE Xplore, www.ieee.org/ieeexplore	Provides access to IEEE reports, journals, transactions, and magazines, IEEE conference proceedings, and current IEEE standards, all published since 1988.

If you've never seen a U.S. patent, look at Figure 8-5. Each front page includes the inventor's name (patentee), owner at date of issuance (assignee), date issued, citations to other relevant patents and articles, one drawing, and an abstract. Following the front page is a disclosure section, in which the inventor describes or "discloses" how his or her invention works and how it relates to or improves on existing solutions to the same problem; and a claims section, in which the inventor gives the legal description of what is actually protected by the patent.

When to Use Patent Information. In your professional work, you might want to do a patent search to do the following:

- Find out about a specific patent
- Learn about recent inventions in a particular field
- Find out if your invention has already been patented
- Gain ideas for further development of your invention
- See what inventions known competitors have patented

```
1824205 NTIS Accession Number: N95-16175/2/XAB
    Simulation of the Coupled Multi-Spacecraft Control Testbed at
the Marshall Space Flight Center
    Ghoah, D. ; Montgomery, R. C.
    National Aeronautics and Space Administration, Hampton, V&.
    Langley Research Center.
    Corp. Source Codes: 019041001; ND210491
    Oct 94 22p
    Languages: English
    The Role of Computers in Research and Development at Langley
    Research Center p. 497-517.
    NTIS Prices: (Order as N95-16453/9, PC A99/MT A06
    Country of Publication: United States
    The capture and berthing of a controlled spacecraft using a
robotic manipulator is an important technology for future space mis-
sions and is presently being considered as a backup option for
direct docking of the Space Shuttle to the Space Station during
assembly missions. The dynamics and control of spacecraft configura-
tions that are manipulator-coupled with each spacecraft having inde-
pendent attitude control systems is not well understood and NASA is
actively involved in both analytic research on this three dimen-
sional control problem for manipulator coupled active spacecraft and
experimental research using a two dimensional ground based facility
at the Marshall Space Flight Center (MSFC). This paper first
describes the MSFC testbed and then describes a two link arm simula-
tor that has been developed to facilitate control theory development
and test planning. The notion of the arms and the payload is con-
trolled by motors located at the shoulder, elbow, and wrist.
    Descriptors: 'Attitude control; •Computerized simulation; 'Con-
trol theory; •Dynamic control; •Manipulators; •Robot arms; •Space
shuttles; •Spacestations; •Spacecraft configurations; •Spacecraft
control; •Spacecraft docking; Equations of motion; Ground tests;
Payloads; Robotics; Shoulders; Space missions; Wrist
    Identifiers: HTISMASA
    Section Headings: 84A (Space Technology—Astronautic«)
```

Figure 8-4 Typical record available from NTIS (National Technical Information Service). Notice that the paper described is part of an internal Langley Research Center report and that the entire report must be purchased (see Order as). Reports labeled with a PC (or price code) can be ordered from NTIS (1-800-336-4700) on paper or microfiche.

Many engineers are unaware of the enormous amount of technical information contained in patent documents. In fact, you cannot find descriptions of most of the technology contained in U.S. patents in any other source. Because patent searching is complex, read about the process in one of these two general sources:

- *General Information Concerning Patents: A Brief Introduction to Patent Matters.* U.S. Department of Commerce, Patent and Trademark Office: Washington, DC: 1992.

Table 8-5 Famous and Not-So-Famous Patents

Coca-Cola Company. *Design of the bottle*. Patent number 696,147

William M. Mirick. *Correction fluid composition*. Patent number 3,674,729

Mervin R. Williams. *Illuminated hula hoop*. Patent number 4,006,556

Lynda S. Samen. *Combined earthquake sensor and night light*. Patent number 4,978,948

Yau, Chiou C., et al. *Ozone-friendly correction fluid*. Patent number 5,199,976

Aaron Harrell. *Pneumatic shoe lacing apparatus*. Patent number 5,205,055

W. Roelofs, et al. *Cockroach attractant*. Patent number 5,296,220

Israel Siegel. *Gravity powered shoe air conditioner*. Patent number 5,375,430

David Falco. *Versatile necktie tying aid gauge*. Patent number 5,505,002

F. Robert Egger. *Bicycle helmet*. Patent number 5,651,145

Dean L. Kamen, et al. *Human mobility vehicle*. Patent number 6,367,817

Put the patent number in the search field at patft.uspto.gov/netahtml/srchnum.htm and take a look.

- Timothy Wheery, *Patent Searching for Librarians and Inventors*. Chicago: American Library Association, 1995. Explains important differences between copyrights, patents, and trademarks.

You can also learn about patents and patent searching at these sites:

- University of Texas, McKinney Engineering Library, Patent Searching Tutorial. www.lib.utexas.edu/engin/patent-tutorial/index.htm.
- Penn State University, Schreyer Business Library's Patent Search Tutorial: www.libraries.psu.edu/instruction/business/Patents/index.html.

How to Find Patent Information. The best places to find recent patent information are the *Official Gazette* of the United States Patent and Trademark Office (USPTO), the USPTO home page, and commercial databases such as Lexis/Nexis or U.S. Patents Fulltext. The *Official Gazette* contains brief descriptions and drawings of the some 1500 patents granted every Tuesday. To get to the USPTO patent-search page, go to www.uspto.gov/patft/index.html.

Because the data is so voluminous, there is also an annual two-volume hardcopy version of the *Official Gazette*, entitled *Index of Patents* and available in many large public and university libraries. A more efficient way to search patents is to go to a Patent and Trademark Depository library (PTDL). A list of these libraries can be found in the *Official Gazette* or on the USPTO's home page (www.uspto.gov/). PTDLs provide free access to other search tools.

For information on patents issued in other countries, the best sources are online. The cost varies widely. Three useful sources are as follows:

JAPIO	Japanese patents
INPADOC	European patents
DERWENT	World patents

US006043842A

United States Patent [19]
Tomasch et al.

[11] **Patent Number:** 6,043,842

[45] **Date of Patent:** Mar. 28, 2000

[54] **REMOTE SURVEILLANCE DEVICE**

[75] Inventors: **Michael D. Tomasch,** Massapequa Park, N.Y.; **Anthony G. Martin,** Trabuco Canyon, Calif.

[73] Assignee: **Olympus America, Inc.,** Melville, N.Y.

[21] Appl. No.: **08/775,311**

[22] Filed: **Dec. 31, 1996**

[51] Int. Cl.^{7} **H04N 7/18**
[52] U.S. Cl. **348/164**; 348/143; 385/118
[58] Field of Search 348/164, 143, 348/65, 68; 128/898; 385/118; 73/864.73; H04N 7/18

[56] **References Cited**

U.S. PATENT DOCUMENTS

Re: 33,572 4/1991 Meyers

4,027,159	5/1977	Bishop
4,261,204	4/1981	Donaldson 73/864.73
4,574,197	3/1986	Kliever
4,696,544	9/1987	Costella 385/118
4,707,595	11/1987	Meyers
4,998,282	3/1991	Shishido 381/77
5,130,527	7/1992	Gramer et al.
5,215,105	6/1993	Kizelshteyn 128/898

Primary Examiner—Howard W. Britton
Attorney, Agent or Firm—Michaelson & Wallace; Peter L. Michaelson; John C. Pokotylo

[57] **ABSTRACT**

A remote surveillance system including an imaging device and an IR light source for surveying a relatively dark area, a remote surveillance system including an imaging device and an insertion tube guide, and an insertion tube guide for receiving an insertion tube of an imaging device.

47 Claims, 12 Drawing Sheets

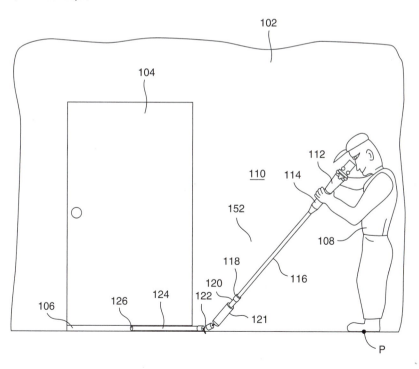

Figure 8-5 Front page of a U.S. patent document. Averaging ten pages, patent documents also include disclosure and claims sections as well as interesting drawings like this one.

If you are interested in applying for a patent for your own work, begin with these two good resources:

- Susan Ardis, *An Introduction to U.S. Patent Searching.* Englewood, CO: Libraries Unlimited, 1990.
- David Pressman, *Patent It Yourself,* 9th ed. Berkeley, CA: Nolo Press, 1995. (Text and software are available.)

PRODUCT LITERATURE

A gold mine of information for engineers can be found in product literature, which includes product, manufacturer, company, and vendor catalogs, as well as product selectors, buyers' guides, and so on. You'll find performance data, photographs or drawings of products, data books for computers and integrated circuit devices, and application notes and other information about specific products. Topics can range from aerospace ordnance equipment to transportation and vehicle equipment or supplies. Sales representatives are most willing to help you get what you want, and most libraries can provide you with company addresses.

When to Use Product Literature. If you are on a design project, product literature is indispensable. You can get the dimensions or performance figures for specific components, accessories, or equipment related to your project. These resources also help if you are producing or marketing your own company's products. Using these resources enables you to know what is already available in your field and to compare currently available products.

Product literature is aimed at selling products and only briefly provides details of specific products. An individual product catalog usually features just one product, whereas manufacturers' catalogs show a variety of products for sale from a specific manufacturer (see Figure 8-6). Vendor catalogs, on the other hand, show all products for sale by the vendor and are designed for fast and easy comparison among several competing brands. These usually provide purchasing information such as the manufacturer's location and phone number, and most give at least limited product specifications, performance data, drawings, test data, and application details.

One example of a product catalog dedicated to a specific field is the *Electronic Engineer's Master* catalog (EEM). Designed to help users see a range of similar products, EEM is a collection of pages from catalogs of companies around the world. It provides an advertiser's index and a manufacturer's index, including local sales offices and distributors.

How to Find Product Literature. To get an idea of the enormous variety of products described in catalogs, look at the annual *Thomas Register of American Manufacturers*, commonly called *ThomCat*. The first 16 volumes provide access to the names, addresses, and telephone numbers of about 150,000 U.S. manufacturing firms. You can look up companies that make a specific product—for example, back-

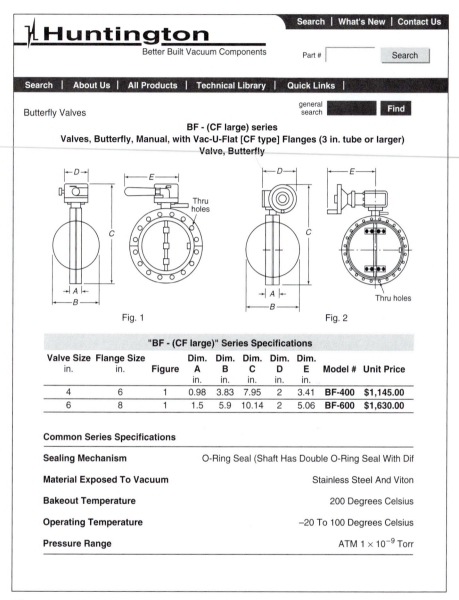

Figure 8-6   Web page from Huntington Laboratories' online catalog. Huntington Mechanical Laboratories, Inc., 2003, www.huntvac.com (reprinted by permission).

hoes. Other volumes of interest include a U.S. tradename index; you can use this index to find out who manufactures Teflon, for example. Volumes 19 to 26 consist of selected pages from individual company catalogs. You can get copies of most catalogs described in *ThomCat* directly from the publisher.

Other useful sources of product information are the large microfilmed collections of manufacturers' and industrial catalogs, such as *Information Handling Service's Visual Search Microfilm Files* and *Information Marketing International*. These consist of copies of complete catalogs along with indexing, permitting you to look up several manufacturers of potentiometric multimeters, for example.

STANDARDS AND SPECIFICATIONS

Most products we use daily are designed and produced in accordance with certain standards or specifications. The length of toothpicks, the softness of toilet paper, and the different grades of sandpaper are all controlled by agreed-upon industrial standards. These standards are essential if you want to be able to consistently fit light bulbs into sockets, screw nuts onto bolts, replace engine parts, or rely on the strength of concrete. Furthermore, we live in a society that is both safety conscious and increasingly alert to the quality of mass-produced consumer items.

When to Consult Standards and Specifications. As a design engineer, you must be aware of what standards, specifications, or codes already exist that might be relevant to your product. One professional engineer puts it this way:

> *You will usually be informed of the applicable specs by your managers, but they may miss some. If you do not comply with an applicable spec, you will have to redesign, with cost to your organization and criticism of yourself whether or not it was your fault. It is good practice to assure yourself independently that you know all applicable specs.*

> Lawrence L. Kamm, *Successful Engineering: A Guide to Achieving Your Career Goals* (New York: McGraw-Hill, 1989), p. 145.

The terms *standards* and *specifications* are often used as synonyms because they can both refer to the guidelines by which something is measured, designed, tested, or manufactured. When a formal distinction is made between the two terms, however, *standard* is more general whereas *specification* is more specific. For example, you might provide specifications for how you want your house built, but the builder will still have to abide by standards set by the city.

The standards for specific products are set by the trade associations, companies, manufacturers, and professional societies involved in those products, and also by government agencies and international standards organizations. Table 8-6 lists a few of the hundreds of organizations that produce standards.

How to Find Standards and Specifications. So many engineering standards exist, and so many different organizations issue them, that you might think finding a

Table 8-6   Some Producers of Standards

American Gear Manufacturers Association	AGMA
American Society of Mechanical Engineers	ASME
American Society for Testing and Materials	ASTM
Institute of Electrical and Electronics Engineers	IEEE
National Wire Rope Manufacturers Association	NRMA
Underwriters Laboratory	UL
American National Standards Institute (U.S. treaty representative to other standards producers)	ANSI
General Services Administration (a main U.S. government generator of standards)	GSA
Department of Defense (another main U.S. government generator of standards)	DOD
German Institute for Standardization (major non-U.S. engineering standards organization)	DIN
International Organization for Standardization (another major non-U.S. engineering standards organization in Geneva, Switzerland)	ISO

standard is like looking for the eye of a needle in a cosmic haystack. Fortunately, this is not the case. Nowadays, engineers have access to efficient ways to locate standards information:

- American National Standards Institute (ANSI). *Catalog of American National Standards* (traditional printed resource), www.ansi.org (website).
- Department of Defense (DOD). *Index of Specifications and Standards* (traditional printed resource).
- Information Handling Service (IHS). *Industry Standards and Engineering Data: Number and Subject Index* (general tool covering multiple organizations).
- National Standards Association. *Standards and Specifications* (focuses on government and industry standards and specifications, including FEDSpecs and MILSpecs.
- International Telecommunications Union. ITU Recommendations (www.itu.ch).
- International Organization for Standardization. ISO Online (www.iso.ch).

Figure 8-7 shows the typical structure of a standard. In this brief example, the resolution is subdivided and involves different organizations. This illustration also demonstrates how standards numbers work. For example, ASTM F807-83 consists of the following: the organization (ASTM), the standard number (F807), and a hyphen followed by the year issued (83). For the validity of this standard, pay attention to the parenthetical "R 1993." This indicates that the issuing organization reviewed and reapproved this 1983 standard in 1993. Always check to be sure you

Resolution:

Used for: Limiting Spatial Resolution: LSR; Resolving Power

See also: Image Intensifiers: Numerical Aperture

Tolerances

ANSI PH3>609-80 Dimensions for Resolution Test Target for Photographic Optics (R 1987) NFP(A) T2.9.6 R1-90. Hydraulic Fluid Power—Calibration Method for Liquid Automatic Particle Counters. (Revision/Re designation ANSI B93.28-1973)

Cathode-Ray Tubes

EIA TEB25-85 Survey of Data-Display CRT Resolution Measurement Techniques.

Copying Machines

ASTM F807-83 Standard practice for Determining Resolution Capability of Office Copiers. (R 1993)

Image Processing Systems

AIIM TR26-93 Resolution as it Relates to Photographic and Electronic imaging.

Photographic Lenses

ANSI PH3.63-74. Method for Determining the Photographic Resolving Power of Photographic Lenses. (R 1991)

Figure 8-7 Description of an industrial standard, from *Industry Standards and Engineering Data: Subject Index.*

are using the most up-to-date standard, unless you are concerned with a historical design problem and need the standard that was current at the time the equipment was designed (or failed).

U.S. Government Specifications. As one of the world's largest buyers of practically every kind of civilian and military product or service, the U.S. government has produced enormous quantities of standards and specifications describing requirements for its purchases. In 1995, a policy decision was made concerning FEDSpecs and MILSpecs. The U.S. Congress and the General Accounting Office (GAO) instructed government agencies to use ASME, ASTM, ANSI, IEEE, UL, and other standards or specifications whenever possible. If you are interested in the specifications for items bought by the General Services Administration and the U.S. Armed Forces, you should begin with the following:

- United States Federal Standards and Specifications (FEDSpecs)
- United States Military Standards and Specifications (MILSpecs)
- Department of Defense Index of Specifications and Standards (DODISS)

Some engineering libraries have industry standards collections containing information on the standards that are to be used by the government and that are formulated by organizations such as IEEE, ANSI, and ASME. Those with large collections may also provide you with access to one of the electronic indexes mentioned earlier. Sometimes you can get help directly from the issuing agency, especially if you are buying a copy of the document. Commercial vendors, such as Global (1-800-624-3974), provide express-service copies of U.S. and international standards to those who need them.

INTERNET RESOURCES FOR ENGINEERING INFORMATION

To this point, we have discussed mostly print-based methods of finding information, some of which have electronic analogs, such as *Engineering Index* online. However, the Internet now offers some powerful alternatives for information research to the engineer.

USENET NEWSGROUPS

Usenet refers to newsgroups on the Internet, which are ongoing Internet-based discussions focused on a particular topic and used by people all over the world. Newsgroup activity is simply a series of email messages, written rapidly by anyone with or without any qualifications, that are all strung together. Despite their questionable reliability and fragmented nature, newsgroups are a good way to get information and opinions and, more importantly, names and addresses to contact.

Just what sorts of groups are out there? How do you find out what newsgroups exist for your field or topic? Usenet newsgroups will never be anything but chaotic, but Table 8-7 gives you some starting points.

Table 8-7   Finding Engineering-Related Usenet Newsgroups

Google Groups, groups.google.com	Visit groups.google.com, click **sci.**, and select, for example, **sci.engr**, and then perhaps **sci.engr.biomed**. When you select the group, the most recent posts are at the top. (Also try **comp.**)
Tile.net, www.tile.net/	For "category," select **news,** and for "Text to search for," enter biomedical to see the same newsgroup as above.

Electronic Mailing Lists

Mailing lists are another Internet tool like Usenet newsgroups, with the same reliability questions and discontinuous nature. Here's how a mailing list works: Let's say you are a civil engineer and want to enter into discussions with other civil engineers worldwide. You subscribe to CIVIL-L: Thereafter, you receive any email that anyone else subscribed to CIVIL-L sends to that list; any email you send to the list gets sent to all the other subscribers to CIVIL-L. Many, but not all, of these electronic mailing lists archive their email activity. Not only can you watch current email for your topic, but you can also search these archives for your topic and see what subscribers have said about it. Table 8-8 lists some examples of mailing lists.

As with Usenet newsgroups, you need to maintain a healthy skepticism about the information you find in electronic mailing lists. Because it takes more effort to subscribe to a list than it does to engage in newsgroup activity, postings on a mailing list are likely to be more considered and professional. But it's still email and it's still the Internet—messages tend to be written hastily, and you cannot be sure of the senders' qualifications. Even so, the information and contacts you can get from following a mailing list or rummaging in its archives can be invaluable, even if you don't trust the information itself.

Table 8-8 Examples of Engineering-Related Electronic Discussion Lists

CAEDS-L	Computer-aided engineering design (CAEDS) interest group.
CHEME-L	Covers the role of chemical engineering in technology and world economies and serves as an open forum for various technical, professional, and educational issues.
CIVIL-L	Civil engineering research and education.
ENVENG-L	For those interested in education, research, and professional practice relating to environmental engineering.
GENTECH	For the exchange of information among concerned scientists, activists of grassroot groups, and other organizations about the impacts of genetic engineering on environment and society.
MATERIALS-L	For those involved in both teaching and research in materials science and engineering. Interested participants might be in materials departments in universities, in other engineering or science departments, or in industry or government research.
MECH-L	For the discussion of mechanical engineering (ME), including finite element methods, composite materials, and other ME-related topics.
METALLURGY-L	Covers all aspects of metallurgical engineering, including (but not necessarily limited to) mineral processing, extractive metallurgy, hydrometallurgy, pyrometallurgy, metals refining, alloying, welding, casting, and metallography.
TDR-L	Discussion of time domain reflectometry issues for engineering and geo measurements.

Table 8-9   Finding Engineering-Related Electronic Mailing Lists

CataList, www.lsoft.com/lists/ listref.html	Official catalog of LISTSERV lists. Browse or search the 72,443 public LISTSERV lists on the Internet, read brief descriptions, or subscribe.
Directory of Scholarly and Professional E-Conferences, www.kovacs.com/directory/	Made available by Kovacs Consulting: Internet & Web Training. Click Search the Directory and enter a search term such as biomedical engineering.
Google, www.google.com	Type "discussion lists" engineering in the Google search field.

How do you find mailing lists in your areas of interest? Try the possibilities listed in Table 8-9. Figure 8-8 shows a sampling of what you'd see if you searched on biomedical at www.kovacs.com/directory. When a discussion list is archived, you can search past email activity for topics and see what others have said. In fact, searching the archives of a discussion list is often better than posting your question to the list.

Directory of Scholarly and Professional E-Conferences

Search results for: biomedical engineering
94 records found

1   Title: IBEGROUP
   Description: Open to all biomedical engineers and technologists. The group's main objectives include making contributions to the science of biomedical instrumentation, providing biomedical engineers with an opportunity to develop interdisciplinary contacts worldwide.
   Subscription: listserv@lists.wayne.edu
   Moderated: No
   Archives: Yes
   Contact: medphys-request@lists.wayne.edu
   Broad: Life Sciences
   Middle: Biology
   Specific1:   Biomedical
   Specific2:   Engineering

2   Title: BIZ-BIOTECH
   Description: The purpose of this discussion list is to bring together businessmen and scientists from industry and academia to work on biomedical and biotechnical projects.
   Subscription: listserv@netcom.com
   Moderated: No
   Archives: Yes
   Contact: Stefan Gruenwald grunwald@netcom.com
   Broad: Life Sciences
   Middle: Biology
   Specific1:   Biomedical
   Specific2:   Business

Figure 8-8   Electronic discussion lists—search samplings. These items were found by searching on biomedical engineering at the Directory of Scholarly and Professional E-Conferences at www.kovacs.com/directory/.

ELECTRONIC NEWSLETTERS AND JOURNALS

Newsletters are less formal publications than e-journals, which report research findings. Professional societies may publish newsletters, some of which are freely available on the Web. There is nothing unique about electronic newsletters and journals in the way that there is with Usenet newsgroups and electronic mailing lists. It's just that increasing numbers of these newsletters and journals are no longer being published in print media and exist in electronic form only. Access to these online-only periodicals varies widely. Some you subscribe to—they come weekly, monthly, or quarterly in your email (some are free; some you pay for)—whereas others you access online from sites where they are posted or archived.

When to Use Electronic Newsletters and Journals. Electronic journals and newsletters provide technical and practical information of interest to professionals, notices of conferences, and ads for related services and products. Keep in mind that some of the sites exist primarily to sell products or services or to promote industry or special interests.

How to Find Engineering-Related Electronic Newsletters and Journals. Many large libraries of institutions offering degrees in engineering maintain lists of newsletters, e-journals, and websites for various disciplines. Table 8-10 lists some places in which to begin your search.

INTERNET SEARCH TOOLS

The Internet world and the information world in general have been revolutionized by the World Wide Web. Almost all of the access and search techniques previously discussed are integrated and consolidated by the Web. Not only can you search, but you also can view—and not just text but also graphics.

ENGINEERING RESOURCES ON THE WEB

With the Web, you are still limited to (or endangered by) whatever people feel like making available on it. Plenty of engineering resources exist on the Web, but it's not like an organized library. Still, there are fascinating resources available, which Table 8-11 helps you access.

WEB SEARCH TOOLS

How can you survey the entire Web for engineering resources related to your topic? Several tools exist for searching the Web: World Wide Web Worm, Lycos,

Table 8-10   Finding Engineering-Related Newsletters

Web search engines, such as Google, www.google.com/	Type the name of the institution in quotes and the word library in the search field, *or* type the words newsletter and engineering in the search field. *Note:* Experiment with truncating search terms (use the root of the word); for example, use engineer*.
EEVL, the Internet Guide to Engineering, Mathematics and Computing, www.eevl.ac.uk/index.htm	Try the Engineering E-journal Search Engine (EESE) at www.eevl.ac.uk/eese/ for access to newsletters and e-journals.
American Society of Civil Engineers, www.asce.org/ American Society of Mechanical Engineers, www.asme.org/ American Academy of Environmental Engineers, www.enviro-engrs.org/ Institute of Industrial Engineers, www.iienet.org/ Society of Automotive Engineers, www.sae.org/ World Coal Institute: *Ecoal*, www.wci-coal.com/	Go to the website of a professional society for your engineering specialization, and search for newsletters and journals there. To the left are some examples.
Engineering companies	These produce printed technical journals, company journals, or newsletters that, in their electronic form, may be made available to everyone.
U.S. Department of Energy, www.doe.gov/	Some government agencies, such as DOE, publish free online newsletters. For an example, go to afdc3.nrel.gov/documents/altfuel news/ and take a look at *Alternative Fuel News*.
Product, industry, nonprofit, or service-oriented websites	These organizations, such as the Solar Electric Power Association, produce newsletters to attract viewers to their sites and to inform members of events and news. For an example, see *The SEPA Record* at www.resourcesaver. com/file/toolmanager/Custom O63C178F42219.pdf.

WebCrawler, the Internet Jumpstation, and Web Nomad. Check any commercial book about the Web for their addresses. Table 8-12 lists some websites that can help you find additional search tools.

Table 8-11   Engineering Resources on the Web

WWW Virtual Library, vlib.org/	Links to information sources in many different fields. Select Engineering for links to civil engineering, mechanical engineering, materials engineering, and more. These in turn link to engineering-related societies, journals, newsletters, design and research centers, academic institutions, special projects, and tips on finding other online engineering information.
Galaxy Search Engine and Directory, www.galaxy.com/	Similar to the WWW Virtual Library in collecting links to engineering resources.
Yahoo!, dir.yahoo.com/science/ engineering/	Similar to the WWW Virtual Library and Galaxy in providing collected links in multiple levels of subcategories.
INFOMINE: Scholarly Internet Resource Collections, infomine.ucr.edu/	Select PhysSc, Engr, CS, and Math and then type engineering (or your specific topic) as the search term.
EEVL: Internet Guide to Engineering, Mathematics, and Computing, www.eevl.ac.uk/	Provides huge annotated, categorized lists of engineering-related resources (based in Great Britain, but fully international in scope).
eFunda: Ultimate Online Reference for Engineers, www.efunda.com /home.cfm	Like an online reference book; provides formulas, mathematics, unit conversions, online calculators, processes, design information—even online job search and resume posting.
EngNet Engineering Directory, www.engnetglobal.com/	Directory with built-in tools, such as unit converters, four main divisions—Engineering Categories, Industry Categories, Brandnames, and Companies— and industry news. Search the site by product, company, or brand name.
CiteSeer (ResearchIndex), citeseer.nj.nec.com/cs	A "scientific literature digital library" for the dissemination of and feedback regarding scientific literature.
GrayLit Network, www.osti.gov /graylit/	"Gray literature" is defined as "foreign or domestic open source material that usually is available through specialized channels and may not enter normal channels or systems of publication, distribution, bibliographic control, or acquisition by book sellers or subscription agents" (U.S. Interagency Gray Literature Working Group, "Gray Information Functional Plan," 18 January 1995). Use GrayLit Network to search the gray literature of U.S. federal agencies over he Internet, without first having to know the sponsoring agency.

(continues)

Table 8-11 (*continued*)   Engineering Resources on the Web

Scirus, www.scirus.com/	Referring to itself as "the search engine for scientists," Scirus allows you to search specifically for conferences in your subject area: Use the advanced search option, choose Conferences from the Information Types list, enter a search term, and Scirus presents you with conferences related to your subject.
National Academies Press, books.nap. edu/v3/makepage.phtml?val1 =subject&val2=eg	Some publishers, such as the National Academies Press, offer free online access to the full text of a small number of recently published books at their websites. Use the URL to the left to find engineering books.
Online Books, onlinebooks. library.upenn.edu/	Provided by the University of Pennsylvania, this resource enables you to find books using their Library of Congress call number and to read the full text online. Engineering titles can be found in the "T" Library of Congress call number section at onlinebooks.library.upenn.edu/webbin/book/ subjectstart?T.

Table 8-12   Web Tools for Finding Web Search Tools

Search Engine Watch, searchenginewatch.com/	Comprehensive site that lists and evaluates Web search engines, provides Web searching tips, and publishes news about searching. The search engine links are in categories such as major search engines and directories and specialty search engines.
Infopeople Search Tools Chart, www.infopeople.org/search/chart.html	Good starting point for finding the appropriate search engine for your needs.
Best Search Engines Quick Guide, www.infopeople.org/search/guide.html	Another good starting point for finding the appropriate search engine for your needs.
Best Subject Directories to Use, www.lib.berkeley.edu/TeachingLib/ Guides/Internet/SubjDirectories.html	Describes several general subject directories and gives tips on finding more specialized ones. This guide is part of the UC Berkeley Teaching Library Internet Workshops series.
Types of Search Tools, www.lib. berkeley.edu/TeachingLib/Guides /Internet/ToolsTables.html	Discusses search engines, directories, searchable database contents, metasearch engines, and gateway pages, with suggestions on how to decide which to use for your particular needs.
Other Internet Search Tools, notess.com/search/others/	Features specialized search tools such as email list directories, tools for searching the "invisible Web" and blogs, free online reference tools, and more.

If the URL for any of these resources has changed, look up "unified search engines" in books on the Web.

EXERCISES

If you are not familiar with library-based information sources, find a technical topic that is of interest to you, and look for information related to it in as many of the following sources as you can.

1. Check the catalog at your library and one of the Internet-accessible libraries mentioned in this chapter. For the three most useful-looking books related to your topic, make a bibliographic entry using the format shown in the documentation section of Chapter 11.

2. Using *Ulrich's International Periodicals Directory* or some other journal-locating resource, find three useful-looking journals related to your topic. Make a bibliographic entry for each one.

3. Using one of the periodical indexes discussed in this chapter, find three useful-looking articles related to your topic in technical journals. Make a bibliographic entry for each one, again using the format shown in Chapter 11.

4. Using NTIS or some similar resource mentioned in the technical reports section of this chapter, find three technical reports related to your topic, and make a bibliographic entry for each.

5. Using one of the patent indexing resources discussed in this chapter, find at least one patent related to your topic, and make a copy of the record that is displayed.

6. Using one of the catalogs described in the product literature section of this chapter, find at least one company involved with products or services related to your topic, and make a bibliographic entry for it.

7. Find one electronic mailing list, one Usenet newsgroup, and one website related to your topic. Use the search tools available on the Internet and the Web to assist you in these searches.

BIBLIOGRAPHY

Ardis, Susan B. *A Guide to the Literature of Electrical and Electronics Engineering*. Littleton, CO: Libraries Unlimited, 1987.

Basch, Reva, and Mary Ellen Bates. *Researching Online for Dummies*, 2nd ed. New York: Wiley, 2000.

Bird, Linda. *The Complete Guide to Using and Understanding the Internet*. New York: Prentice-Hall, 2003.

Doherty, Paul. *Cyberplaces: The Internet Guide for Architects, Engineers, Contractors and Facility Managers*. Kingston, MA: R.S. Means, 2000.

Fischer, Martin A. *Engineering Specifications Writing Guide*. Englewood Cliffs, NJ: Prentice-Hall, 1983.

Kamm, Lawrence L. *Successful Engineering: A Guide to Achieving Your Career Goals*. New York: McGraw-Hill, 1989.

Lo, Jack, and David Pressman. *How to Make Patent Drawings Yourself: Prepare Formal Drawings Required by the U.S. Patent Office.* Berkeley, CA: Nolo Press, 2002.

Lord, Charles R. *Guide to Information Sources in Engineering.* Englewood, CO: Libraries Unlimited, 2000.

Neville, Tina M., Deborah B. Henry, and Bruce Neville. *Science and Technology Research: Writing Strategies for Students.* Lanhan, MD: Scarecrow Press, 2002.

Pressman, David. *Patent It Yourself.* Berkeley, CA: Nolo Press, 2001.

Thomas, Brian J. *The World Wide Web for Scientists and Engineers: A Complete Reference for Navigating, Researching, and Publishing Online.* Bellingham, WA: SPIE Press, 1998.

9

ENGINEERING YOUR PRESENTATIONS

The skills required to present complex technical information in a personal, pro-fessional, and accessible manner are invaluable in today's information-saturated society.

> Laura Gurak, *Oral Presentations for Technical Com-munication* (Boston: Allyn & Bacon, 2000), p. xix

Engineers are often called on to speak, and many engineers find they speak a lot. Whether you give an impromptu five-minute briefing or a formal one-hour presenta-tion at a technical seminar (or something in between), you should see your talk as a great opportunity to share information and to show that you know how to communi-cate. Few of us are naturally gifted with such capabilities, and some of us are almost petrified at the thought of talking before a group, but the skills possessed by good speakers can be learned. The principles discussed in this chapter will enable you to become a confident that speaker that people will listen to because you transfer infor-mation efficiently and effectively—that is, with a minimum of noise.

PREPARING THE PRESENTATION

Developing a worthwhile presentation is like developing a product: Research and planning are crucial in the early stages. We all know what it's like to have to come up with a spontaneous briefing or unexpected oral report, but fortunately we usually have some lead-in time before we talk. Using that time to work through the procedures that follow will help you design a successful presentation.

ANALYZE YOUR AUDIENCE

Much of what was said at the beginning of Chapter 2 about focusing on your reader and purpose *before* writing can be applied to preparing for an oral presentation. We've all been bored by talks that were over our heads, too simplistic, or unrelated to our interests. Don Christiansen, a former editor and publisher of *IEEE Spectrum*, recounts one of his early experiences:

> *As a young engineer, I was invited to address an IEEE Section meeting. My subject was an unusual stereophonic/quadraphonic audio system developed at CBS Laboratories. This technical presentation may have been my first before a large engineering audience. I worried at the prospect. I prepared and projected a number of slides containing a bunch of mathematics that no one could follow during a brief exposure. After all, I had sat through many conference papers that were ritually peppered with unintelligible (at least to me) equations. I had responded in kind, despite my audience having many spouses present—most of whom hadn't a clue what their mates did for a living. I was grateful to the wives, who did not boo or stamp their feet, but discreetly nodded off.*

> Donald Christiansen, "Engineers Can't Write? Sez Who!"
> *IEEE-USA News & Views* (June 2003), p. 4.

To make sure you don't do the same thing to others, ask yourself the questions in Table 9-1 when beginning to prepare your talk, and make sure you have as clear an idea of the answers as possible.

Table 9-1   Some Questions to Ask About Your Listeners *Before* You Talk

---

- Who will the key individuals in my audience be?
- What needs or concerns do they have regarding my topic?
- What are *my* objectives for this talk?
- How knowledgeable are my listeners about my subject?
- How can I get their attention and interest right away—and keep it?
- What are their attitudes likely to be regarding what I have to say?
- Do I need to work on changing their attitudes, and if so, what is the best way to go about it?
- What benefits are they going to get from listening to me?
- What kinds of questions are they likely to ask?
- What kind of feedback do I want?

DECIDE ON YOUR PRIMARY PURPOSE

Your purpose in talking is intimately related to the makeup of your listeners and the reason they are sitting in front of you. Are they there for instruction, information, insight, to be persuaded, or what? What action or change do you feel they need to undertake? Knowing exactly what kind of assignment you have will also determine your foremost purpose. Engineering presentations can take many forms, as Figure 9-1 indicates, each with a specific purpose and organizational requirements.

Make sure you know what you are getting into, what is expected of you, who your audience is going to be, and what you want to accomplish by talking to them. Decide exactly what you want your listeners to take away from your talk. Then you will be on solid ground while preparing the remaining features of your presentation.

DETERMINE YOUR TIME FRAME

The cardinal rule here is *never* to speak longer than you are supposed to. To avoid annoying busy people or offending speakers who come after you, check how much time you have been allotted. Knowing your time limit will also help you decide how much detail you can go into, how much time you should allow for questions or discussion, and how much time you can spend on an introduction and conclusion or recommendations if you have some.

As Figure 9-2 illustrates, how deeply you go into different aspects of a typical engineering topic is related to how much time you have to speak. The tops of the pyramids in the figure represent the least you could say on a topic—perhaps a single

Figure 9-1 Just a few of the many kinds of presentations engineers find themselves giving.

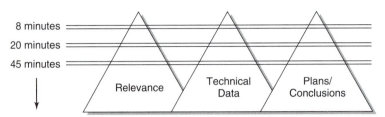

Figure 9-2   No matter how little time you have to talk, you can say something worthwhile in an engineering presentation. You just have to correlate how deeply you delve with the time that you have.

sentence—while the true bases of the pyramids (unseen in the illustration) represent everything that could possibly be said. This is perhaps why we almost always impose time limits on speakers; otherwise, they might go on forever!

It's been claimed that any subject can be covered in virtually any amount of time. A speaker could compress the creation of the universe into three or four sentences—or fewer—if necessary. In the same way, an expert could talk for hours (probably to a rather small audience) on the mating habits of the Gambian giant pouched rat. Whatever you do, don't decide that since you have a lot to say in a short time, you should speak as fast as possible while rapidly clicking slides on the screen. You will just as rapidly lose your listeners.

IDENTIFY YOUR KEY POINTS

As indicated earlier, oral presentations normally don't permit us to give all the minis-cule details about every aspect of our topic. So don't expect to say everything that could possibly be said on your subject. With a sharp awareness of your main purpose and time frame, decide *what the most important points are* that you want to get across to your audience and how you want to develop those points in the time you have.

You may need to be quite mathematical about this. If you have 20 minutes to make four important points, you might subtract the time you want for an introduction and conclusion and divide what is left by four, thus leaving from 3 to 4 minutes for each point. In some presentations you may not want to give equal time to each point. For instance, if you have to discuss five reasons why a project should be canceled or why your company should invest in new equipment, you might determine which points will meet the most resistance from your audience. Then aim to spend more time explaining those points while giving briefer treatment to the less controversial ones.

CHOOSE AN ORGANIZATIONAL PLAN

Your subject and purpose, and to some extent the time you have, will help determine how to organize your material. Most presentations can roughly be broken down into an introduction, the main points, a conclusion, and a question and answer period. Table 9-2 lists ways to organize your central material. Some presentations will call for combining some of these organizational plans, of course.

PREPARE AN OUTLINE AND NOTES

Writing an outline and notes helps you clarify in your own mind how best to present your material. They also give you a means of deciding how much time to allot to each

Table 9-2   Examples of Ways You Can Organize Your Material for an Effective Oral Presentation

Organize Your Material in This Format . . .	If, for Example You Want to . . .
1. Time sequence	Describe progress on a project or steps in a procedure Relate decisions leading up to an action or occurrences that led to a problem
2. Spatial sequence	Describe a piece of equipment or a physical area such as a test site or plant facilities Outline a physical process
3. Decreasing importance	Give your most important points first, down to the least important (e.g., relating six ways to improve or prevent a situation)
4. Increasing importance	Work up to your most important point (e.g., presenting some minor reasons for an action, change, or decision, followed by the major reasons)
5. General to specific	Present a general point followed by specific examples (e.g., "We've got to improve production," followed by concrete ways to do so)
6. Specific to general	Be persuasive (e.g., citing examples of personal injury to lead to the point that more stringent safety regulations are needed at your plant)
7. Comparative	Compare and contrast equipment, approaches, or ideas on such aspects as costs, durability, reliability, ease of operation, etc.
8. Familiarity	Begin with the familiar first, leading your audience into an understanding of the unfamiliar (e.g., talking about corporate needs or problems)
9. Difficulty	Present data in order of the easiest first and progressing to the hardest, as in a training session or tutorial
10. Controversiality	Begin with the least controversial points in order to be diplomatic about sensitive issues (e.g., why changes should be made in a project in which people have some ego investment)

point, and will be helpful when you rehearse the presentation. However, extensively relying on an outline or notes during the actual presentation can be dangerous, because you will give the impression that you don't know your topic thoroughly.

Your notes and outline may range from a few hastily scribbled ideas—if that much—jotted down a few moments before an unexpected briefing, to a complete manuscript of every word you intend to say. Reading a word-for-word written version of your talk in front of an actual audience is *not* a good idea, however, unless you feel very insecure or are giving a highly technical conference paper calling for extreme precision and accuracy. Even then, few people want to sit while a paper is read to them; after all, they could read it themselves in the comfort of their own homes or offices.

While preparing your presentation, determine which prompts will best keep you on track when giving it. The main cues engineers tend to use are the following:

- An *outline* of the complete talk, with key ideas highlighted or in large print to be quickly glanced at if necessary as the presentation goes along (see Figure 9-3).
- *Note cards* numbered in the order they will be used, with key ideas and facts clearly written on them. If these are relied on too much, however, you will appear unsure of your material.

> 5. Future **Needs**
>
>   A. Sharper Images
>
>   B. Cheaper Cyclotrons
>
> 6. **Safety**
>
> 7. **Cost** of Current Technology
>
>   Combination with MRI
>
> 8. Is It **Profitable?**
>
>   To **Market Now?**
>
>   For Research?
>
> 9. **Company's Interest**
>
> 10. **Role** of the Company
>
> 11. **RECOMMENDATIONS**

Figure 9-3 Example of part of a possible outline for an oral presentation. Note that key points are highlighted. In an actual outline, you might want to spread each item farther apart to be easily seen with a quick glance while you talk.

- *Visual aids*, such as transparencies or slides. If you are really on top of your topic, your visuals themselves will be all you need to keep on track. They may in fact be the outline of your talk. If you wish, you can make printed copies of them for your own use, with notes written to yourself that can be quickly referred to if needed.

CREATE SUPPORTING GRAPHICS

Because we live in an increasingly visual age, and because people remember information better when they both hear it and see it, most effective engineering speakers support their talks with illustrations of some kind. Graphics also work to your advantage because preparing them forces you to organize and rehearse your presentation and possibly discover weak spots that need attention. Showing them will save you time during the presentation because you won't have to write the information down on a board or flip chart. They can also serve as cues for you, reminding you of what you want to cover and the order in which you want to cover it. You should at a minimum plan to use visuals wherever you feel they will

- Simplify a point
- Clarify a point
- Stress a point
- Show critical relationships between ideas or facts

Channels for Graphic Support. For engineers, one of the most common means of showing graphics has traditionally been the overhead projector with *transparencies* (also known as foils, overheads, visuals, or view graphs) that you prepare in advance. Transparencies have the advantage of letting you add information on them with a wax marker while showing them, or even using additional ones as overlays. If you have several transparencies, remember they can be slippery. It's embarrassing to see them slide off the table and across the floor, so you might want to consider a matting of some sort for each one. *Flip charts* are useful for ongoing illustrations or emphasis during your talk, as is a *chalkboard*. Slides shown from a *slide projector* can be effective also, especially if you can do a professional job with color and artwork— but this can be time-consuming.

The most popular way to display visuals nowadays is from a laptop *computer* connected to a projector, showing the audience what you have created on a graphics program such as PowerPoint, Lotus Freelance, or Adobe Persuasion. Presentation software such as these allow you to progress through your talk by calling up graphics through your keyboard or a wireless mouse. Such programs also allow you to add numerous features to your presentation, such as sound, zooming text, and colorful templates. The danger here is the temptation to get too fancy and to try to dazzle your audience with your artistic skills rather than by presenting clear, visually accessible information. Furthermore, the most impressive visuals you can make will not lessen your need to speak clearly, effectively, and enthusiastically.

If you are talking about a specific piece of equipment that you can bring into the room with you, do so—as long as your audience is small and close enough to be able to see it. If you hand it around, remember to get it back. This might seem obvious, but in the afterglow of a good presentation, with people crowding around you with praise or questions, it's easy to forget to retrieve your widget.

Designing Your Graphics. Any good graphics program will give you everything you need to create graphs, pie charts, bar charts, flow charts, and so forth. Nowadays, with only a few hours' training any engineer can produce almost any kind of graphic with programs such as PowerPoint. Some specialized programs allow you to create excellent illustrations of equations, electrical circuits, and other technical data. With the growing use of scanners, you can now copy and present professional illustrations or photographs of just about anything. If you show scanned material in your presentation, be aware of any copyright restrictions that might apply, and give credit (usually at the bottom of your slide in small print) to the source of any such material. Chapter 11 of this book deals with the question of citing your sources.

As we implied earlier, the most dazzling transparency or slide will impress no one if the information it contains is not easily accessible. In fact, anything you put on the screen that cannot readily be grasped by your audience—because it's either too complex or too small—is worthless. This point is particularly worth heeding if only because we see it so often ignored by engineering speakers.

- *Too complex:* Don't let your visuals suffer from *information overload.* Each should be as simple as possible, portraying the bare facts—you can always elaborate verbally. Even quite technical material can be reduced to manageable concepts on a screen. For example, something as complex as an electronics circuit can be broken up into constituent parts after a simplified overview has been given, as shown in Figures 9-4 to 9-6.
- *Too small:* If your visuals consist of lists or other written information, *make the words easy to read* (see Figure 9-7). This means using at least a 24-point font size, preferably boldface. It's best to have no more than eight lines of print on a slide or transparency, and better not to use all capital letters because they make for harder reading. A page of text or an illustration photocopied from a book or journal rarely makes a good overhead.

When you present written information on the screen, don't crowd it. Provide ample margins and plenty of white space between and around the lines. You might want to use bullets, checks, or other marks to emphasize points, but resist the temptation to go overboard with the variety of fonts and clipart now available. Don't let your artwork overwhelm the information on the screen.

PREPARE HANDOUTS

Think carefully about whether you want to provide handouts, and if so, what kind and when you will hand them out. Many speakers wisely avoid handouts altogether,

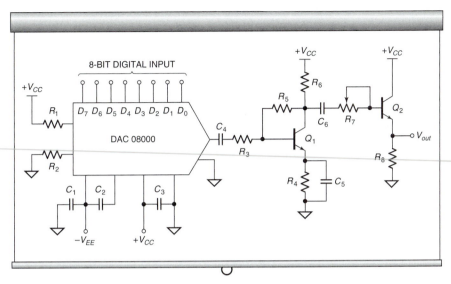

Figure 9-4   An example of an overcrowded transparency. Far too much information is thrust upon the audience here. One way to make this material more accessible would be to reduce the circuit first to a block diagram, as shown in Figure 9-5, and then, if more detail is needed, to expand the drawing one block at a time on separate visuals, as in Figure 9-6.

as they feel they distract their audience's attention. On the other hand, some speakers pass out copies of their overheads or slides, often reduced in size, so listeners can make notes on them. Distributing an outline of your presentation may be a good idea,

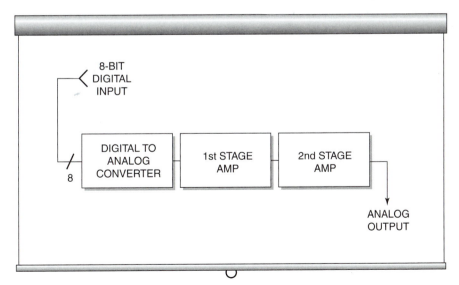

Figure 9-5   Simplified block diagram of the circuit in Figure 9-4.

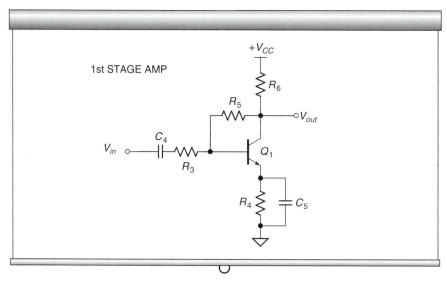

Figure 9-6 The center block from the diagram expanded to part of the original circuit.

especially if the topic is detailed and covers a lot of material, but the choice should be yours. Sometimes you might need to provide supporting evidence for your talk, such as samples, brochures, or other data. Plenty of successful speakers, however, expect their listeners to focus solely on the presentation itself and to take their own notes as they wish.

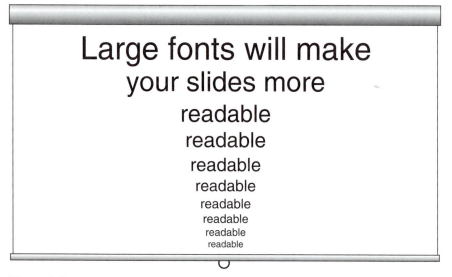

Figure 9-7 Use at least 24-point print on your overheads if you want your audience to read them.

The dilemma with handouts is *when to pass them out.* If they're distributed at the wrong time, they will almost certainly distract from your talk because people tend to look at what is given them right away, and may ignore you or only partially listen. You need to decide beforehand when the best time to distribute material will be, and how and when you will refer to it, so that it adds to your listeners' concentration and understanding rather than takes away from it.

PREPARE YOUR INTRODUCTION

When thinking of how to begin your talk, remember these points:

1. Your audience may be asking themselves, at least subconsciously, "Why do I need to hear this?" or "Why should I be here right now?"
2. Your audience has a limited attention span.

To help solve both problems, design your introduction to *let your audience know right away what your topic is and of what benefit it is to them.* Then provide a sense of direction by giving an overview of where you're going in the presentation and what you plan to cover. Let your audience know how long you intend to speak if it's not already known. Many speakers lose their audience right away because they fail to follow these procedures at the outset of their presentation.

PREPARE YOUR CONCLUSION

Design the end of your presentation to focus the audience's attention solely on essentials. Depending on the type of presentation you give, you will be able to reinforce your message by, for example,

- Summarizing what you have discussed
- Stressing your central idea once more
- Reviewing your key points
- Restating your recommendations or decisions

An appropriate final slide can be a great help here. Above all, give your audience a lasting impression of what you want them to take away from your talk, such as the feeling that you have solved a problem or concern or have provided new insights. Don't suddenly stop talking at this point, however. Make a note to close gracefully with something like "And this concludes my presentation. Thank you for your attention. Are there any questions?"

GET READY FOR QUESTIONS

If there is going to be a question period after your talk, spend some time during the preparation stage to anticipate and get ready for them. Put yourself in the place of your listeners: Are they likely to find any part of your talk especially difficult, detailed, or controversial? Are they likely to hold any opposing viewpoints? Are there areas you may not be able to go into as thoroughly as you would like, due to time restraints, and which might therefore generate questions? What could be the "worst" question asked? Also, can you think of diplomatic ways to encourage questions from people who are reluctant to ask? (Sometimes a friendly smile or "I'd *love* to have some questions" is all that's needed.)

Remind yourself at this point that when questions come, it's often a good idea to repeat them aloud in some form before answering. The repetition is useful for people who didn't hear the question clearly in the first place, and the delay might give you time to gather your thoughts.

PRACTICE, PRACTICE, PRACTICE

Keep in mind these letters: **PPPPPP** (plentiful practice prevents painfully poor presentations). Some speakers go over their material—outline, notes, visuals—up to seven times before presenting it; for others, this would be overkill. Most speakers rehearse at least twice, however, if their talk is of any significance or if they feel unsure of their material.

Depending on the importance of your talk, you may decide to have at least one dress rehearsal if you can, preferably in the room where you'll be presenting. This will let you get familiar with the room and any equipment to be used. An audio or videotape of this rehearsal, if possible, would enable you to self-critique your performance. On the other hand, you might want to find a trial audience to listen to your first run and give you some feedback. Friends, colleagues, a spouse—or even yourself in the bathroom mirror—can be good audiences to practice on.

Perhaps the most valuable outcome of careful practice is the self-confidence you gain. One antidote to nervousness about speaking in front of a group is to be able to walk into that room knowing you're completely in control of your subject and ready to present it in an effective manner—confidence you can only gain by first practicing as much as possible.

DELIVERING THE PRESENTATION

When was the last time that you sat through two and a half hours of a scientific presentation and wished that it would go longer?

Michel Alley, *The Craft of Scientific Presentations*
(New York: Springer-Verlag, 2003), p. vii

All your preparation efforts are aimed at one goal: to give an effective, noise-free presentation that will produce the desired results. By the time you stand in front of your audience, you should have fixed many of the potential glitches that can surface in oral presentations. Knowing your subject and audience makeup will have helped you determine the information you need and how you need to communicate it.

Most engineers can prepare a presentation well enough given a little awareness, analysis, and preparation time, yet the sad fact remains that plenty of lackluster and somewhat boring presentations occur every day in business and industry. As with a written report, such presentations can be greatly improved by the elimination of noise.

AVOIDING NOISE IN ENGINEERING PRESENTATIONS

In an oral presentation, noise can be defined as anything that prevents the message from effectively getting into the minds of the audience. Following are some causes of noise that frequently occur in engineering (and other) presentations.

- *Speaking too softly.* A common problem with beginning speakers is a tendency to speak too softly. Ironically, such softness is a form of noise because it prevents the message from clearly getting through to the listeners. You don't want to blast your audience out, but you do want everyone to hear you. Try to project your voice relative to the room and audience size. If some listeners can't hear you, you're wasting their time.

- *Speaking too slowly or rapidly.* A slow, labored pace with too many pauses causes boredom and decreases your credibility. The reverse of this is to talk too rapidly, either because of nervousness or because you're so much more familiar with your topic than your audience is. Aim for a normal conversational speed, but remember that pausing and deliberately slowing down once in a while can help you stress important points.

- *Speaking monotonously.* You may have heard stories of the dull college professor who dreamed he was giving a lecture only to wake up and find he was. How you talk often makes a bigger impression than flashy visuals and what you say combined. You could be explaining the never before revealed secrets of time travel and yet find few paying attention if you sound bored to death. Hypnotic monotony can be avoided by varying your pace and your pitch—by speaking the way most people do in lively and energetic conversation. *Enthusiasm* on your part will encourage your audience to listen to you.

- *Using verbal fillers.* When a speaker needs to pause or is uncertain of what to say next, irritating and empty catchwords or phrases like *uh, umm, basically,* and *yu'no* sometimes take over. Don't distract your audience with a high UPM (umms per minute) rate. In informal conversation, the word *like* seems to have become almost epidemic: "Like, I'd be happy to help, like, but I don't have enough, like, time." Try to avoid this kind of noise in your presentations. There's nothing wrong with being silent for a few moments while gathering your thoughts.

- *Becoming a statue, pendulum, or traveler.* Because you are to a considerable extent a visual yourself when you give a presentation, you need to be aware of the effect you may have on your audience through your movements. Some speakers tend to freeze physically when in front of a group, and to remain in that position for the entire talk. Others like to sway back and forth without moving their feet. Both are distracting to an audience and do not add liveliness to the presentation. The swinging can even produce a hypnotic effect. Try for a natural stance and movements when in front of your listeners, with some foot movement but not enough to wear out your audience as they follow you back and forth across the room.

- *Blocking the screen.* Too many engineers stand directly in front of the screen and stay there throughout their talk, thus frustrating their neck-craning audience and wasting any effort put into their graphics (see Figure 9-8). It's just as bad to stand partially off to the side yet still block the screen for those sitting at the sides. If possible, before your audience arrives, have a colleague sit in various seats and let you know where you should avoid standing for long. If you don't have time for this luxury, move around enough during your presentation to avoid blocking anyone's view continuously. Better yet, stand far enough to the side to prevent screen blockage from ever becoming a problem.

- *Reading from the screen.* Generally avoid reading your slides during a presentation—they are aids but are not meant to be your entire talk. Straight read-

Figure 9-8   You invest a lot of planning and work into your visuals, so don't create noise by standing between them and your audience.

ing takes time away from a deeper explanation of your topic, and may bore your listeners because they can read for themselves. Also, be wary of dimming the lights to make your visuals easier to read. Low lights can make people drowsy and can hide facial expressions and the eye contact you need to have with your listeners.

STRENGTHENING YOUR PRESENTATIONS

Use an Informal Style. When making an engineering presentation, you're not delivering a sermon (usually) or pronouncing on a profound legal intricacy. Generally, the best style is an informal one, paralleling as closely as possible the normal conversational mode of everyday life. It's quite all right to use contractions (*it's, don't, couldn't,* etc.), even if you avoid them in formal writing. Using pronouns such as *you, your, I,* and *we* will help you relate to your audience's interests and needs, and will indicate you are interested in them as people rather than as an impersonal mass. Avoid long, complex sentences, substandard grammar ("Me and Jim here aren't no experts, but . . ."), and any technical jargon not readily understood by your listeners.

Make Clear Transitions. You may have a well-organized talk full of important details, but you could still lose or confuse your audience if you don't show the connections between your ideas. The key is to be explicit. Tell your audience when you're moving on to another aspect of your subject or are about to give an example. Your visuals will assist you, of course, but make your dialogue as user friendly as possible. Keep your listeners in the picture by emphasizing connections and transitions in your thinking by using simple words and phrases like these:

First	On the other hand
Next	As you can see
To begin with	For example
Initially	Also
Furthermore	Finally
Consequently	In conclusion
As a result	To sum up

When you overlook such transitions in a written report, your readers can at least go back over the material a few times and try to figure out the connections for themselves, exasperating as that might be. Someone listening to a talk has no such opportunity.

Repeat Key Points. No matter how brief the presentation, you're going to have at least one main point you want your audience to go away with. Don't be afraid to repeat yourself—your listeners need to know what you consider the most important aspects of your subject.

Given the sad fact that even the best of speakers may have someone in the audience whose attention strays, there is a lot to be said for that old piece of advice "tell your listeners what you're going to tell them, then tell them, and finally tell them what you have told them." As we pointed out earlier in the section on preparing conclusions, it's essential that you repeat your key points in a concluding summary.

Use a Pointer. A pointer is the best way to focus your audience's attention on your key points while you explain what they're looking at. The *laser pen* projects a red dot onto the screen and can be aimed from anywhere in a room. Its drawbacks are that you could permanently injure someone's eye by directing the beam at them, and that an unsteady hand will cause the laser's dot to dart around surprisingly. Hold the laser pointer firmly pressed to your side when aiming if you are at all nervous.

Note Some people are unable to see a small laser dot on a screen.

A *straight metal* or *wooden stick pointer* is always available, but you have to stand fairly close to the screen to use it. This limitation can overly restrict your movement and may also cause you to block the view for some people. The *retractable stick pointer* is also easily available but has the same potential drawbacks. If you use a stick pointer, hold it with the arm closest to the screen so you don't have to turn away from the audience every time you point (see Figures 9-9 and 9-10). Be careful not to overuse it, to hit the screen loudly with it, or to unconsciously wave it around when not pointing. Some speakers have a distracting tendency to open and close this kind of pointer repeatedly because of nervousness, or to even scratch themselves

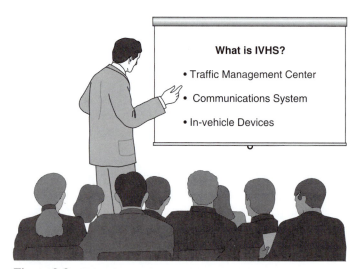

Figure 9-9 Using the arm farthest from the screen to point pulls you from eye contact with your audience.

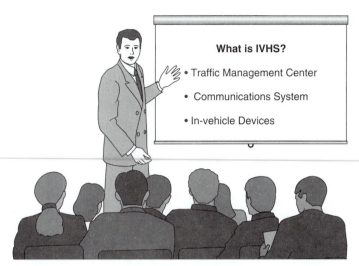

Figure 9-10   Using the arm closest to the screen allows you to talk while facing your listeners.

with it. When not actually using it, keep your pointer firmly clasped in one or both hands, and resist the temptation to conduct an imagined symphony with it.

Maintain Eye Contact. You increase your credibility a great deal by looking at your audience as you talk. Whereas avoiding eye contact could give the impression that you're shifty or unprepared, looking at your audience helps establish rapport with each member in a small group and creates a sense of intimacy with a larger group.

Start off by making eye contact with the friendliest faces and the most attentive people in the audience. As you progress in your talk, try to hold visual contact with each person for a few seconds and then move on to someone else. Looking at individuals also enables you to pick up feedback on how they are receiving your message; puzzled looks or frantic note taking, for example, might show that you need to go back over something or slow down. You may even ask the audience if you should repeat an explanation or complex idea.

Be Ready for Unexpected Questions. You can prepare all day for questions and still land at least one question you never dreamed of. Try not to appear surprised or defensive when this happens—you've prepared a lot and know your subject well. Two strategies for tackling unexpected questions are as follows:

1. Simply say you don't know. People will respect you for being honest, and you can still offer to supply possible sources for the information later.

2. Offer to talk with the questioner after your presentation. This may be the best answer if (a) the question is too involved for the discussion you are in, (b) the

question is not really related to your topic, or (c) the question is hostile and you don't want to get into an argument. Rarely will anyone seek you out afterward unless they are genuinely interested in information you may have.

Accept Your Nervousness. Although this fact may not be very encouraging if you're just starting to give presentations, the best cure for nervousness is experience. Until you have this experience, accept your nervousness as perfectly normal. We all suffer from it (although our nervousness is often much less noticeable to others than we might think). Even the most experienced speakers sometimes get tense before facing an audience; some learn to use their anxiety as a positive energizing power that helps them to be more alert and lively. If you have problems with stage fright, consider the following tips:

- Enter the presentation room knowing you've worked hard on your presentation and have practiced delivering it. In other words, give yourself as much reason as possible *beforehand* to be confident of your knowledge and ability. Then try to concentrate on your topic rather than on yourself. Appropriate gestures and facial expressions will often occur naturally in a well-prepared speech.

- Take some deep breaths before entering the room. Even a short walk around the building or a few simple physical exercises may help relieve anxiety. You can't continue with these activities once you're in the room, of course, and they can never substitute for the self-assurance that comes with really knowing your material.

- Try to have a few friends or colleagues in the audience who can give you moral support through a reassuring smile or nod of the head. If everyone in the room is a stranger, look for a friendly face or two and exchange a few words of banter before launching into your presentation. This can relieve quite a bit of the tension you may be feeling.

TEAM PRESENTATIONS

Because teams of engineers frequently collaborate on a project, compile a proposal, or report on a new product, you are likely to be involved in team presentations. These allow individuals to speak in turn on a topic, each with his or her own "part." Group presentations also permit a specific aspect of a complex subject to be presented by the individual who worked on it rather than by a team spokesperson. This kind of presentation is additionally effective because

- Teamwork reduces everyone's preparation workload.
- Longer presentations are possible without exhausting one person.

- Speakers can enjoy team support during the presentation.
- The variety of speakers helps hold the audience's attention.
- Each topic can be explained (and questions answered) expertly.

PREPARING FOR A TEAM PRESENTATION

Whether giving the presentation with one partner or several, the first step is to get together to analyze your audience and purpose. Then decide on the main points to be stressed, the order in which the material will be covered, and who will cover which topic. The team leader should clearly partition the topics and make sure that each speaker sticks to the assigned topic and doesn't cover any other speaker's material.

It's essential to allocate time to each speaker early in the process so that everyone can prepare accordingly and the presentation can conform to any required time limits. Decide beforehand whether questions from the audience should wait until after the entire presentation or should follow each speaker (if you have these options).

Because collaborating engineers do a lot of communicating with each other by rather impersonal memos or telephone messages, meeting to prepare the presentation may also provide an opportunity for all presenters to get to know each other better. If group members are familiar with each other, the presentation is likely to be more relaxed and look more polished and professional.

SHARING THE PRESENTATION

Assigning different parts of the presentation to alternate speakers prevents a long presentation from becoming monotonous, arouses more audience interest, and provides clear structure to the presentation. Plus, it's nice for each speaker to have a breather.

To ensure that your group presentation flows well, pay particular attention to how you are going to move from one speaker to the next. A simple lead-in like "And now Eva is going to cover the financial aspects of the project" might be all you need. If there are just two of you, break down the topic so you can "leapfrog," with each person eventually speaking for approximately the same total amount of time.

MAKING A DRY RUN

Groups, like individual speakers, need to put aside plenty of time for practice. Rehearsals will uncover any information gaps or neglected subtopics and will ensure that all parts of your presentation are carefully merged. If team members have never performed together before, practice will be essential to ensure coordination in the presentation. As an individual team member, you will also need to be certain you can conform to time limits: There is no better way to make enemies than to dominate the presentation and speak longer than you should, shrinking fellow speakers' time.

Checklist for an Oral Presentation

The items in Figure 9-11 can be used to evaluate a presentation. You can also use the list *before* you give your talk. Put yourself in the place of your audience and try to get a sense of how they would grade you as they listen to your presentation. You might even get a friend to check you out on each item during a dry run.

INTRODUCTION
- ☐ Creates favorable atmosphere
- ☐ Creates appropriate pace
- ☐ Hooks listener's attention
- ☐ Relates subject to listeners
- ☐ Presents clear central idea

BODY
- ☐ Reveals careful audience analysis
- ☐ Supports central idea
- ☐ Maintains audience interest
- ☐ Provides technical accuracy
- ☐ Organizes details effectively
- ☐ Allocates time carefully
- ☐ Provides clear transitions

CONCLUSION
- ☐ Ties presentation together
- ☐ Restates central idea
- ☐ Proposes action or response
- ☐ Invites discussion or questions

ATTENTION TO TIME LIMITS
- ☐ Too short
- ☐ Just right
- ☐ Too long

VISUAL AIDS
- ☐ Are clear and easy to read
- ☐ Look professional
- ☐ Avoid information overload
- ☐ Clearly support related ideas
- ☐ Are enough

DELIVERY

Sound
- ☐ Clear volume and pronunciation
- ☐ Effective diction
- ☐ Varied speech patterns
- ☐ Absence of *uh-huh, y'know, basically* . . .
- ☐ Adequate enthusiasm
- ☐ Standard grammar and usage
- ☐ Good question response

Appearance
- ☐ Professional posture and appearance
- ☐ Appropriate gestures and mannerisms
- ☐ Effective use of pointer
- ☐ Consistent eye contact with audience
- ☐ Competent handling of notes and visuals

Figure 9-11 The aspects of an effective oral presentation. Even if they don't seem important to you, they will to your audience.

LISTENING TO A PRESENTATION

You will probably listen to more presentations than you give during your engineering career, yet listening is the most neglected of all communication skills. You may have already found there's nothing more frustrating than working long and hard on a presentation just to have an unresponsive or uninterested audience, and the point can be made that the responsibility for a successful presentation lies partly with the audience.

Here are some ways to be a good listener:

- Maintain natural eye contact with the speaker.
- Show by your posture that you are alert, interested, and well disposed toward the speaker.
- As much as possible, ignore distractions such as people talking or other external noise.
- Take notes on the speaker's most important points.
- Develop at least one question in your mind, and ask it at the appropriate time.

In fact, one of the best ways to be a good listener is to ask questions. A question lets the speaker know you're paying attention and that the presentation has made you think. Also, a question helps a presenter decide which information is most important and where the main concerns of the audience lie. Even if you understand everything presented, why not ask about something you found particularly interesting or new?

It's not hard to see how actively focusing on a speaker and concentrating on what is being said establishes a sense of empathy that leads to more efficient information transfer. Being an active listener, then, is not just a matter of being kind to the speaker; it also rewards you with a more complete appreciation, knowledge, and evaluation of the material presented.

EXERCISES

1. Ask one or two engineers what kinds of oral presentations they make on a fairly regular basis. How long do they talk each time? What do they talk about? Who is their audience? How do they meet the needs of the audience? Do they use any graphic aids or handouts while presenting? What kinds of feedback do they get? How do they know whether their presentation has been successful?

2. Listen to someone giving an oral presentation and evaluate his or her performance as best you can using the checklist in Figure 9-11. (You may need to be very diplomatic about this.) After evaluating the presentation, think of ways you might improve it if you had to give it yourself.

3. Take a written report—your own, if possible—and turn it into an oral presentation. What do you have to consider when preparing the talk that was not important when writing the document? Do you have to change the order or emphasis of any material? Do you find material in the written report that can be presented graphically in the talk? Does an outline of the report give you ideas of what to present graphically? How will you introduce and conclude your presentation? Does the fact that you may be asked questions after the talk cause you to think differently about your material than you did when writing the report?

4. Think of the various people you have listened to, such as teachers, fellow professionals, preachers, or politicians. What are some of the best things you have heard or seen these

people do? What are some of the worst? Which were the most effective speakers? Which were the most ineffective? What can you learn from both kinds that will help you improve your own skills in giving oral presentations?

BIBLIOGRAPHY

Alley, Michael. *The Craft of Scientific Presentations*. New York: Springer-Verlag, 2003.

Beebe, Steven A., and Susan J. Beebe. *Public Speaking: An Audience-Centered Approach*, 4th ed. Boston: Allyn & Bacon, 2002.

Beer, David F., ed. *Writing and Speaking in the Technology Professions,* 2nd ed. New York: IEEE Press, 2003.

Emery, Blake, and Karen Klamm. "Effective Listening," *Proceedings of the 33rd International Technical Communication Conference*, pp. 129–131, Detroit, 1986.

Gurak, Laura J. *Oral Presentations for Technical Communication*. Boston: Allyn & Bacon, 2000.

Woelfle, Robert M., ed. *A New Guide for Better Technical Presentations*. New York: IEEE Press, 1992.

10

WRITING TO GET AN ENGINEERING JOB

The resume is your main vehicle for presenting yourself to a potential employer. The central question to ask in preparing your resume is, "If you were an employer, would you want to read this resume?" . . . Visual impact and appearance are extremely important.

> Raymond Landis, *Studying Engineering: A Road Map to a Rewarding Career*, 2nd ed. (Discovery Press, 2000), p. 223

The key to resume writing excellence is in presenting it the right way. Most people make the error of just listing their experience and qualifications; this ends up being a rather boring document. A good resume should not only demonstrate your skills and experience, but should also give the reader a good indication of the type of person you are. It needs to have personality.

> Engineers International, "Preparing the CV," http://www.engineers-international.com/careerscv.html, accessed July 16, 2003

Two tools commonly used to seek employment are the resume and the application letter.[1] You send one or both of these to prospective employers when you are in a job search. The combination depends on the potential employer—some request only the resume, some request only the letter, and some don't indicate. When you're not sure, send both.

---

[1] Our special thanks to Randy Schrecengost, P.E., for his reviews and recommendations on this chapter.

HOW TO WRITE AN ENGINEERING RESUME

A resume is a summary of your professional experience, education, and other background relevant to the employment opportunity you are seeking. Think of it as highlights of who you are professionally—a summary of your career to date.

The key to writing an effective resume—one that highlights your best qualifications—is a design that can be scanned in about 20 to 30 seconds. Even within that short amount of time, the prospective employer should be able to glance through your resume and still have a decent understanding of your background and qualifications.

CONTINUOUS REDESIGN AND UPDATE

Developing a resume is not a one-time effort. Consider it a work-in-progress: You may have to revise it for every new employment opportunity you seek; you must update it at every accomplishment, milestone, or new phase in your career.

It used to be that a resume was a fairly permanent record of your background, which you updated only infrequently. You could use roughly the same resume for many different job searches over a number of years. However, with increasing competition in the job market and with the availability of desktop publishing software, all that has changed. Now, many job seekers redesign their resumes for every new employment opportunity.

This constant updating is just as important if you settle into one company for a long time. It's easy to forget details about what you've done professionally over the space of just five years. For that reason, keep a working draft of your resume always at hand—whether as a computer file or as a hardcopy printout on which you scribble notes whenever your career takes a new turn. As of 2003, the joke is that you must keep your updated resume on a diskette in your coat pocket at all times.

Note   Be aware that most employers now handle resumes primarily by computer—electronically and online. This automated process creates both new opportunities and new hazards. See "Electronic Resumes" later in this chapter and Roger Munger, "Technical Communicators Beware: The Next Generation of High-Tech Recruiting Methods," *IEEE Transactions on Professional Communication* 45, no. 4 (December 2002).

DESIGN COMPONENTS

The design of a resume is certainly important to success in an employment search. But a resume can't do it alone—many other elements have to be present, such as connections, timing, need, and of course your actual qualifications. Still, a well-designed resume does a number of things for you: It highlights your best qualifications, makes it easy for readers to see them quickly, and conveys polished professionalism that reflects positively upon you.

Chronological or Functional Organization. One of the first issues in resume design is whether to organize your background information chronologically or functionally. To get a sense of these two organizational approaches to resumes, look at Figure 10-1 for a schematic view and Figure 10-2 for a full-content view.

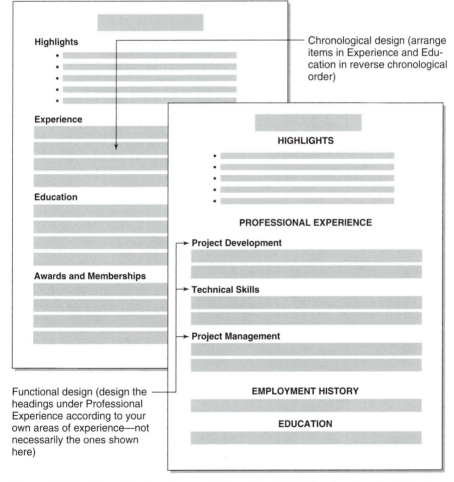

Figure 10-1 Schematic view of resume designs. Decide whether the chronological or the functional design works best for you. Visualize the headings you'll use and their relation to each other and to the body text.

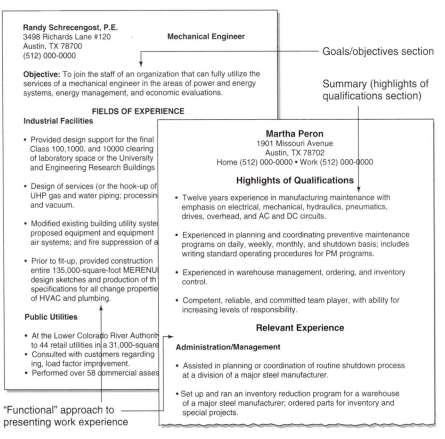

Figure 10-2   Special sections in resumes: the summary or highlights of qualifications section and the goals and objectives section. Using a highlights section that lists your key qualifications allows a potential employer to get a quick picture of who you are professionally. Use the objectives section to indicate your professional focus.

The *chronological approach* divides your background into education, work experience, and possibly military experience (although military experience can be distributed into the education and experience sections instead).

One of the strengths of the chronological design is that it shows your work history—in particular, your responsibilities and projects for each organization you've worked for. In the education section, this design shows where you studied and what you studied while there. However, the chronological design does not give a capsule picture of your key qualifications and your key strengths—that information is spread across work and education sections. (One way to solve this problem is to add a highlights section, discussed later in this chapter.)

The *functional approach* divides your background into groups of related education and experience. For example, you may have taken courses in college on project management, attended professional seminars on the subject, taken lead roles in the management of several projects, and maybe even won an award for your management of a project. All of this could be summarized under the heading "Project Management" in a functionally organized resume (see the schematic illustration of this in Figure 10-1).

The great strength of the functional approach is that it consolidates information about your key qualifications, summarizing all relevant work experience and education for each one. Prospective employers looking for someone with project planning and management experience can quickly discern from the functional approach whether you have what they are looking for.

Of course, the weakness of the functional design is the strength of the chronological design: In the functional design, it is immediately clear where, how, and when you gained your experience or education. However, the chronology of your career is unclear. A solution to this problem is to include a list of your experience and education—no descriptions, just the names and dates (this is schematically illustrated in Figure 10-1 in the headings "EMPLOYMENT HISTORY" and "EDUCATION").

Note   If you are at the beginning of your career, or only a few years into it, consider using the chronological design.

Highlights Section.   Another issue is whether to include a highlights section (called different things, including Summary of Experience, Summary, Highlights of Experience, Summary of Qualifications, Synopsis of Qualifications, Professional Expertise, Qualifications, and so on). This section is popular, particularly for professionals who are several years into their careers. It is particularly helpful in resumes that use the chronological design. In the chronological design, key points about your experience and education are scattered—bits and pieces are cited throughout the work experience and education sections. Readers have to reconstruct your highlights for themselves.

With the highlights section, however, you do that reconstruction for your readers. The summary provides a neat bulleted list of your key accomplishments, key areas of expertise, key education and training, and so on. Even a reader who did not look further in your resume would still get a good picture of who you are professionally.

Notice in Figure 10-2 that a bulleted list is used to make the items in the highlights section more scannable. You position this section just at that point where the eye first makes contact with the page. Many believe that our initial glance makes contact with a page one-fourth to one-third of the way down the page, not at the very top. If you believe that, then putting your very "best stuff" at that point in the resume makes a lot of sense.

Objectives Section.   Still another issue in resume design is whether to include an objectives section. This section describes your career and professional focus. It can indicate the type of work you want to do, the type of position you seek, the type of organization you want to work for, or some combination of these or other objectives.

This section should be brief—no more than two to three lines. It should also be rather specific and not a patchwork of "sweet nothings." For example, avoid stating your objective as "Seeking a challenging, rewarding career with a dynamic upscale company where I will have ample room for professional and personal growth" (as opposed to what?). Instead, try for something specific:

Construction engineer seeking position in HVAC design and energy calculation for residential and commercial structures.

Some experts argue against the objectives section, fearing that it can narrow your opportunities. However, crafty types rewrite this section to correlate with each position they seek. If it's a large, big-city corporation or a small, rural company specializing in a particular technology, then corresponding words are in the objectives section.

Memberships and Licenses. Another important section in engineering resumes is the list of professional organizations and licenses. In a section like this, indicate that, for example, you are a member of the American Society of Mechanical Engineers.

Specialized Equipment and Knowledge. Many engineers also include in their resumes a section that itemizes their technical knowledge. For example, computer specialists may list the hardware and software they know. Electrical engineers may list their skills in such areas as analog circuit and signal analysis as well as digital and control systems.

Miscellaneous Sections. There are many other possibilities for special sections you can include in a resume. For example, if you've published articles in professional journals, create a publications section. If you've received honors and awards, create a section for that. If you have received patents, list those in their own section. If you have various security clearances, list them. The idea is to design the resume so that it emphasizes your best and most important qualifications. Special sections with their own identifying headings are one way of doing that.

Personal Sections. Some, but certainly not all, resume writers include a section at the very end in which they cite loosely relevant personal details about themselves such as interests, nonwork activities, hobbies, memberships, other languages, and so on. Strictly speaking, this sort of information is out of place in the resume—what does the fact that you raise orchids have to do with your career as a structural engineer? Viewed more broadly, however, this kind of information rounds you out as a human being. It gives the prospective employer something to chat with you about while waiting for the elevator—to fill those moments of otherwise uncomfortable silence.

Presentation of Details. In addition to planning the overall design and contents of your resume, you must also decide on how you want to present the actual details of your background and qualifications.

As Figure 10-3 illustrates, there are many ways to show your experience. You can present it as a simple paragraph, as the lowest of the three examples in the figure shows; you can present it in bulleted lists, as the other examples in the figure show. You can highlight the name of the organization you worked for by presenting it first in all caps, bold, italics, or bold italics. Or, you can highlight your title or position by presenting it first, as the rightmost example in Figure 10-3 does.

As to the kinds of details you can present in these sections, there are many possibilities, as shown in the following lists. However, remember to be selective; don't include everything in these lists. Don't bury your best qualifications in a mass of less important detail.

For the experience section, consider including these details:

- Name of the organization where you worked, its address, and phone number
- Your job title and your specific responsibilities
- Brief description of the organization—its products, services, and technical aspects
- Your major achievements, important projects, promotions, and awards
- Experience with technologies, equipment, and technical processes
- Dates of employment with the organization

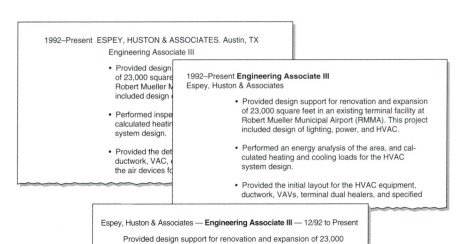

Figure 10-3   Examples of detail formats. Use combinations of list or paragraph format and italics, bold, or all caps in the design of the four main elements: date, organization name, job title, and details.

For the education section, here are some ideas:

- Name and address of the educational institution
- Brief description of the educational institution
- Your major and minors and grade point averages (overall and in your major)
- Major emphasis of study
- Important courses taken
- Brief descriptions of those courses
- Experience with technologies, equipment, and technical processes
- Important projects
- Awards and memberships
- Dates of enrollment and graduation

When you present these details, be as specific as you reasonably can: Cite specific product names, specific measurements and dimensions, specific processes and activities. For example, instead of saying that process improvements that you designed resulted in "considerable savings," say that they resulted in "an average cost savings of $315,000 annually." Instead of stating that your work was done to military standards, state the specific standard, for example, "SAMSO-STD-77-7." Instead of stating that you "redesigned processors for modems," state that you "redesigned Cy-6000 low-gate processors for QAM/QPSK/FSK-mode modems." Generalities are less noticeable than specifics; they have far less impact than specifics; and they seem less real, less authentic.

Also, use strong action verbs when you discuss your background and qualifications. Verbs such as *designed, developed, utilized, coordinated, supervised,* and so on are more striking and memorable than *was involved with, handled,* or *was responsible for.*

OVERALL FORMAT

Figure 10-4 provides a schematic view of some common ways to design the overall format of resumes. You can see that headings can be centered, they can be placed on the left margin but run into the text, or they can be put in their own column separate from the text. Some resumes add ruled lines horizontally or even vertically to increase the visual separation of the components of a resume. (See Figure 10-5 for a full-content example of an engineering resume.)

Format of Headings and Margins. As you design a resume, consider headings and text in relation to those headings. Many resumes use a "hanging-head" design in which the headings are on the far left margin and the body text of the resume is indented about 1 to 2 inches. This design makes the line length of body text shorter and more easily scannable, the headings more visible, and the sections of the resume more visually distinct.

Resume Length and Headers for Multiple-Page Resumes. How long your resume should be depends on how much detail there is in your background and qualifica-

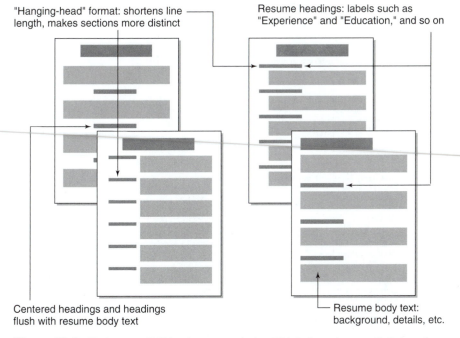

"Hanging-head" format: shortens line length, makes sections more distinct

Resume headings: labels such as "Experience" and "Education," and so on

Centered headings and headings flush with resume body text

Resume body text: background, details, etc.

Figure 10-4 Various possibilities for resume design. Think about the overall design of your resume first—how the headings are positioned in relation to the text; visualize it in blocks like these, without the words, to get an overall sense of the design.

tions. It's likely that early in your career you'll have trouble filling up a single page. After a few years, however, you'll have trouble keeping your resume to one page, then two pages, and so on. The chief problem with long resumes is that prospective employers may not read them as closely as short ones. If you can somehow cut the length of your resume from three pages to one, the prospective employer is more likely to notice and remember your key qualifications. Some resume experts maintain that you should plan for one page of resume for every ten years of experience. However, there are plenty of reasons why this rule of thumb might not be applicable.

In any case, if your resume is more than one page, place headers at the top of the following pages. Design the header to contain some combination of your name, date, and the page number, as in the example shown here:

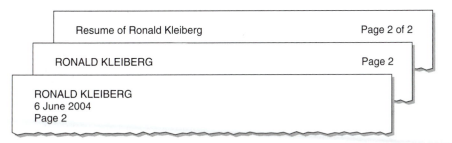

Resume of Ronald Kleiberg   Page 2 of 2

RONALD KLEIBERG   Page 2

RONALD KLEIBERG
6 June 2004
Page 2

Malcolm E. Hunter, P.E.   Page 2 of 2
                        EDUCATION

University of the East   1973 to 1978
Manila, Philippines
DEGREE: BACHELOR OF SCIENCE IN ELECTRICAL ENGINEERING
PROJECT: National Airspace System—Air Route Traffic Control Center
Power System Upgrade.

                        OR
Institute of Electrical and Ele
Electronics Engineers
POSITION: MEMBER

                        PROF
• Engineering analys
• Electrical and I & C
  petrochemical pliat
• Startup and test of
• Project engineer
• Construction engin
• Safety review, test

                        PRO
• Applied Protective
• Power Circuit Anal
• Electrical System
• Electrical Delivery
• National Air Syster
  Panel
• Engine Generator
• Critical Power Con
• Power Control Mor

   LICENSES AND F
• Level II Electrical E
• Quality Safety Mar
• Design Verifier pla

                        REFER
               Available o

Malcolm E. Hunter, P.E.
3498 Oak Shadows #120
Austin, TX 78733
(512) 653-4664

Objective: Senior or supervisory position that utilizes a background in
Electrical Engineering.

PROFESSIONAL EXPERIENCE

Exide Electronics Corporation   May 2003 to Present
Raleigh, North Carolina

POSITION: SITE PROJECT ENGINEER. As a Lead Project Engineer for the
Power System Upgrade for the Air Route Traffic Control Power
Center, responsibilities include supervision of engineering and
design of low- and medium-voltage power distribution systems,
including equipment selection and installation; preparation of
construction drawings; system analyses; coordination of construction and testing activities.

PROJECT: National Airspace System—Air Route Traffic Control Center
Power System Upgrade.

Enercon Services, Inc.   Nov. 2002 to May 2003
Wilmington, North Carolina

POSITION: SENIOR ELECTRICAL ENGINEER. Responsibilities included
engineering evaluation of work requests on various electrical
and I&C systems including technical review, safety review per
10CFR50.59 and design verification per ANSI N45.2.11 for
Carolina Power & Light Company.

PROJECT: Brunswick Steam Electric Plant—Units 1 and 2, Southport,
North Carolina.

Sargent & Lundy Engineers   Mar. 2001 to Nov. 2002
Chicago, Illinois

POSITION: SENIOR ELECTRICAL ENGINEER. Responsibilities included project
engineering, design, and analyses for various electrical and
I&C systems such as generators, HVAC, control rooms, PLC,
auxiliary power systems, station blackout systems, electrical
heat tracing systems, conveyor motor upgrade, high voltage
testing facility, fly ash silo, and blower house for various fossil,
nuclear, and industrial facilities.

PROJECT: S&C Electric High-Voltage Test Facility (Chicago, Illinois), Bryon
& Braidwood Stations, Units 1&2; Labadie Power.

Figure 10-5   Excerpts from the resume of an experienced professional engineer. Notice
the use of small caps for position titles (such as "Site Project Engineer"). The headings on
page 2 of this resume are "Education," "Organizations," "Professional Training," "Proficien-
cies," "Licenses and Professional Certification," and "References."

Table 10-1 on the next page presents some general tips regarding resumes.

Table 10-1   Tips on Writing Resumes

- Don't omit normal words such as articles (*a, an, the*). Make your writing style compact but not unintelligible.

- When referring to your own work, omit *I*. Instead of writing "I supervised a team of 12 designers...," write "Supervised a team of 12 designers...."

- Present education and work experience in reverse chronological order.

- Omit details on age, marital status, gender, religion, handicaps, and other personal matters. Don't include a photograph of yourself.

- Include specific details about qualifications and background: specific product names, specific dimensions, specific processes, and specific technologies.

- Use strong action verbs when presenting details about qualifications and background: *designed, developed, coordinated, supervised,* and so on.

- Make sure that the different sections of your resume are distinct from each other; use spacing, ruled lines, and headings.

- Avoid lengthy paragraphs; keep paragraphs under four lines.

- Indicate the meanings of abbreviations or acronyms—don't assume the whole world knows what "GPA" or the construction "3.5/4.0" means. Spell out the names of organizations; briefly explain their functions.

- Use format consistently: If you present the details of your work experience using one format, use that same format in other similar areas of your resume.

- Use consistent margins. Typically resumes have two to three levels of indentation (for example, in the hanging-head format)—one at which headings align, and another at which text aligns. Make sure all text uses these two levels of indentation and none others.

- Use special typography moderately—for example, bold, italics, underlining, type sizes, and different fonts. Don't go wild with multiple fonts (Times, Helvetica, Thames, etc.) and font faces (bold, italics, underscores, etc.).

- Use special typography consistently: For example, if you put company names in bold in the work experience section, put college or university names in bold in the education section.

- Keep resumes as short as possible—for example, one page at the start of your career; two pages after you've gained substantial professional work experience.

- Keep the resume from spilling over by just a few lines to a second or third page. Force the resume to fill the pages it occupies.

- If your resume is more than one page, put a header on the second and following pages.

- If you send photocopies of your resume, get a high-quality photocopy. Request that high-quality paper be used.

ELECTRONIC RESUMES

You are probably aware that since the late 1990s employers have been increasingly using various electronic and online methods for selecting job candidates. This trend has created much uncertainty, but here are some suggestions.

- *In-house resume-scanning and -searching applications.* Initially, employers electronically scanned hardcopy resumes and keyword-searched them for job candidates. In this early phase, resume writers were warned to use a plain, no-frills font and avoid special typographical effects such as bold, italics, alternative fonts, and different type sizes because of the limitations of scanning equipment. Scanners, however, have improved considerably. Even so, many candidates are expected to send their resumes by email attachment (in electronic form), which eliminates the scanning step altogether.[2]

- *Job boards and websites.* Another popular option involves *job boards*, websites at which job seekers post their resumes for a small fee and at which employers search for candidates for another fee. These can be general, such as Monster.com or America's Job Bank (ajb.com), or specialized, such as the National Writers Union Job Hotline (nwu.org) or EngineersforHire.com.

  Both Internet-based job boards and resume databases enable employers to search for likely candidates by using keywords. For example, if the prospective employer were looking for someone with experience in HVAC layout and calculation of heating and cooling loads, these words as well as close synonyms must appear in the resume.

- *Corporate websites for recruitment.* As Roger Munger explains in his article on high-tech recruiting methods, employers have not remained satisfied with the methods described in the preceding items and have established their own corporate recruitment websites. The practice has become widespread: For example, well over 90% of companies in the manufacturing, healthcare, and high-tech sectors rely on this method. At these websites, employers can tailor and fine-tune the online application process to help them better identify candidates who not only use the right keywords but who also actually qualify for the for jobs they seek.

- *Online profiles.* Munger points out that employers may gradually abandon the traditional resume and direct applicants to fill out questionnaires. Applicants' answers will then be used to construct searchable profiles, which provide more detail, a tighter match with employer requirements, and a consistent format.

In the face of all this variability:

- Assume that your resume will be read, archived, and searched by computers—even for jobs in small companies.

[2]Roger Munger, "Technical Communicators Beware: The Next Generation of High-Tech Recruiting Methods," *IEEE Transactions on Professional Communication* 45, no. 4 (December 2002).

- Carefully follow the application guidelines stated by a potential employer.
- Make sure you send your resume in the expected format: hardcopy printout or electronic file; formatted or plain-text ASCII.
- Include a cover letter if requested by the employer, even if you submit your resume electronically.
- Make sure you format your resume so that it scans properly. Do some test scans yourself on your own resumes.
- Make sure your resume contains keywords relating to your qualifications, to the specific job you are seeking, or both. Use industry-standard keywords.
- Use professional resources—such as journals, newsletters, and conferences—to stay abreast of the evolution of online job searching and online job recruitment.

HOW TO WRITE AN APPLICATION LETTER

Often accompanying the resume is the application letter, sometimes called a cover letter. This letter is the first thing that potential employers see when they open the envelope—the application letter on top, with the attached resume beneath it.

As mentioned earlier, whether to include an application letter with your resume depends on what the potential employer specifies. Sometimes, only the resume is requested; sometimes, only the letter. Sometimes, after prospective employers make their initial selection of candidates, they request the other of the two components.

There are two categories of application letters, based on the information they contain:

- *Cover letters.* In the true cover letter, you simply announce that a resume is attached, indicate that you are investigating an employment opportunity, and specify the position you seek. As illustrated in Figure 10-6, this is a brief letter, the body text totaling less than ten lines. If the job advertisement asks for resumes only, you can still include this type of letter to identify the position.
- *Full application letters.* In the true application letter, you discuss your background and qualifications as relevant to the position you are seeking. The job of this letter is to promote yourself—to highlight the reasons why you are right for the position. This type of letter is the focus of the rest of this chapter.

Which to use? The cover letter is certainly easier to write, but it doesn't do anything for you. The full application letter acts as your proxy, showing the prospective employer specifically which of your qualifications makes you right for the job. If a full application letter is expected, sending only a cover letter makes you seem non-committal or even indifferent.

6307 Marshall Lane
Austin, TX 78703

25 May 2005

Ms. Juanita Jones
Hughes & Gano, Inc.
P.O. Box 1113
Austin, TX 00000

Dear Ms. Jones:

Please accept the attached resume as my application for the position of Process Engineer currently available with your company.

I'll be looking forward to meeting with you at your earliest convenience. I can be reached at (512) 471-4991 during regular working hours or at (512) 471-8691 in the evenings and weekends.

Please contact me if you need any further information about my background or qualifications.

Sincerely,

Patrick H. McMurrey

Patrick H. McMurrey
Encl.: resume

Figure 10-6   Cover letter: a brief correspondence that identifies the position being sought and the purpose of the correspondence. For most job searches, use the full application letter, as described in this chapter.

CONTENTS AND ORGANIZATION OF THE APPLICATION LETTER

The function of the application letter is to introduce you to the prospective employer, state the purpose of the letter (to seek an interview), identify the position you're inquiring about, and summarize your relevant background and qualifications. This last function is the most important. The application letter is not just another form of the resume—it is a careful selection from the resume. It makes a strong case for you as a good candidate for the specific position, primarily by pointing out aspects of your background that are a good fit with the specific job you are seeking.

The following discussion of the contents and organization used in application letters illustrates what's typically done—not the one and only right way.

First Paragraph. The first paragraph of the letter should be brief and do some combination of several things: state the purpose of the letter (to inquire about employment); state how you found out about the opening, if applicable; say something that will catch your readers' attention and make them want to continue reading—and that's it. Keep this first paragraph short, five lines at the maximum.

In this first paragraph, consider using one of several common tricks to catch your readers' attention:

- State something specific about your qualifications that makes you the right candidate for the position.
- Cite information about the company to which you are applying—information that shows you are informed and that relates to the position you seek.
- If possible, mention the name of someone in the company who knows you and can speak knowledgeably—or, better yet, favorably—about you.
- Say something enthusiastic or energetic about the kind of work you want to do, the kind of organization you want to work for, or maybe something about your professional goals.

Whichever of these strategies you use, remember to keep it short. Also, remember to write in terms of the potential employer's perspective. For example, employers don't want to hear at length about your personal goals—only enough to see that you fit in with their operations.

Middle Paragraphs. The middle portion of the letter discusses your qualifications that relate specifically to the employment you seek. Somewhere in these paragraphs, suggest that readers see the attached resume for more detail. In these paragraphs, use the same kinds of organization as in the resume: chronological or functional.

- *Chronological approach.* Discuss your education in one or more paragraphs, and then your work experience in another set of paragraphs. If work experience is your best "stuff," put it before education. (If your education is "old history," just leave it out.)
- *Functional approach.* Focus on the important areas in your qualifications—for example, project management, research and development, quality control, vendor coordination, documentation, and so on. Ideally, reserve a separate paragraph for each of these areas. In each, discuss anything in your background, whether work experience or education, that relates to that area.

These organizational approaches are schematically illustrated in Figure 10-7. If the middle paragraphs take up too many lines, consider using a bulleted list (as illustrated in Figure 10-8). This format enables you to present important details in a condensed, more readable and scannable way.

Final Paragraphs. In the final portion of the letter, you wrap it up: Mention that the resume is enclosed if you've not already done so; urge the prospective employer to get in touch; facilitate arrangements for an interview; and find some parting encouraging, enthusiastic comment to make, such as your strong interest in the employ-

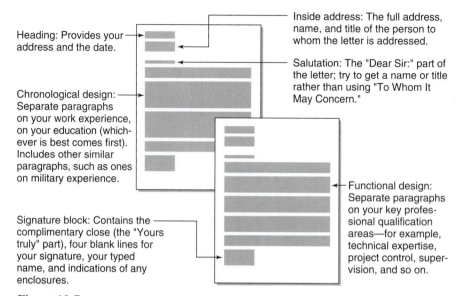

Figure 10-7   Common sections of application letters. You can organize the letter chronologically or functionally, the same as you can the resume.

ment, the company, the profession, and so on. Some job seekers indicate that they will call the prospective employer on a certain date (for example, a week after mailing the letter). Although some might find this tactic too aggressive, it certainly puts pressure on the employer to take action.

FORMAT OF APPLICATION LETTERS

As for the format of the application letter:

- Use a standard business-letter format, such as the one shown in the examples in this chapter. (See Chapter 4 for a discussion of the style and format of business letters in general.)

- Single space the individual components (never double space). Double space between the components.

- Leave four blank lines between the complimentary close and your typed name, and sign your name in that space.

- Do not indent the first lines of paragraphs of the body text. Use standard left and right margins—1 inch, 1.5 inches, or 2 inches are all acceptable. Use wider margins when your letter seems too skimpy.

- Use standard top and bottom margins: The letter can begin anywhere from 1 inch to 3 inches from the top edge of the page; it should end no closer than 1 inch to the bottom edge.

Patrick H. McMurrey
1108 West 29
Austin, TX 78703
(512) 471-9229 (home) (512) 878-6556 (work)

May 25, 2004

Director of Personnel
Automation Associates
7805 Peart Creek Drive
Austin, Texas 78706

Dear Director of Personnel:

I would appreciate your time in evaluating my qualifications in relation to your current needs for a Senior Electrical Engineer in Automation Associates's large building design projects. Attached is a copy of my current resume.

I have over 15 years experience in various facets of electrical design and engineering. Specifically, I have experience in power and control design including analyses for power generation; low-, medium-, and high-voltage power distribution systems; fire detection and protection systems; plant security systems, programmable logic controllers (PLCs); as well as equipment layout for various types of industrial facilities.

I am currently employed with Exide Electronics Corporation as Site Project Engineer in the National Air-Space Federal Systems Engineering Division. My current responsibilities are as follows:

- Lead project engineer for the power system upgrade of Denver, Albuquerque, Indianapolis, and Jacksonville Air Route Traffic Control Centers (ARTCC). This power upgrade is part of Exide's current contract with the U.S. Air Force and the Federal Aviation Administration.

- Supervision of varying numbers of electrical engineers and designers in various engineering tasks for projects such as low- and medium-voltage power distribution system engineering and design.

- Vendor interface for installation of equipment such as diesel generators, switchgears, and power control monitoring systems.

This current work and past projects, along with the references I am including in this letter, all attest to my solid record of initiative, responsibility, creativity, and professional dedication. I am an effective, contributing member of any organization that I am associated with. If you are interested in discussing my experience and capabilities further, please contact me at one of the numbers shown above.

Sincerely,

Patrick H. Mc Murrey

Patrick H. McMurrey
Encl.: resume, reference list

Figure 10-8 Example of an application letter. Notice how much specific detail the writer packs in concerning his experience. Notice also how the bulleted list relieves some of the density of the letter.

- Carefully position your letter on the page. If your letter is short, don't leave it crammed at the top of the page. Use the variables of margins and spacing between text components to position the text of the letter in the upper middle of the page.

- Avoid dense paragraphs. Don't expect readers to labor through paragraphs over 10 or 12 lines long. Use paragraph breaks and numbered or bulleted lists to loosen up dense paragraphs.

- For additional eye appeal, consider creating an attractive, professional-looking design for your name and address, such as those illustrated in Figure 10-9.

Tone in Application Letters. In an application letter, tone may be the most important characteristic, but also the hardest to define. Tone should reflect your view of yourself and the type of professional you want to be. You may want to avoid sounding brash, arrogant, or overconfident—unless that's your personality. The following examples explore how bad tone can result from good intentions.

- *Stiff, overly formal, overly reserved.* When you write an application letter, the pressure is on—obviously. One tendency is to freeze up and be overly cautious. Ironically, this can sound like indifference or create a stiff, reticent, overly formal tone, suggesting a personality that prospective employers would rather avoid.

- *Intimidatingly qualified, or even overqualified.* Tone can also go bad when you overemphasize your qualifications and make yourself sound like a miracle worker. Employers may get uneasy, fearing the prospect of a co-worker with an overdeveloped ego. They may worry about the safety of their own jobs, or they may wonder whether you're stretching the truth or simply lying.

- *Unctuous, fawning, flattering.* It's possible to try too hard to sound bright, positive, enthusiastic, and eager; it's possible to sound phony in saying nice things about the prospective employer.

- *Hungry, desperate.* Avoid the tone that says "I'll do anything!" The anxiously eager tone can go bad when it starts sounding desperate for a job—any job. Maintain your professional focus and integrity: You won't do just any kind of work; you want the kind of work you have trained for.

- *Overly humble, overly simple, above it all.* It might be tempting to adopt the attitude that says "this is who I am, this is what I can do, this is what I have done—take it or leave it." It's simple, humble, plain, no-nonsense. But it can sound so excessively (even aggressively) humble that employers may decide the job seeker will prove unbearably superior, even arrogant.

Bad tone can start from good intentions: Most job seekers want to be cautious and respectful, to show what's good about themselves, to be enthusiastic and complimentary, to sound comfortable and confident professionally, to demonstrate that they are earnest about the employment opportunity, and to be honest and straightfor-

Jane A. McMurrey
801 East 31st Apt. 101
Austin, TX 78701

June 6, 2004

Director of Personnel
Dow Chemical U.S.A.
2020 W.H.D.C. Building
Midland, MI 48674

Dear Mr. Ian Hanson:

I am writing in regards to t
Commercial development
experience, my education
me for this opportunity.

Working at DuPont for five
positions has given me thi
employees. As a Compute
technical and 100 hourly e
communications through t
continuously analyzed and

Working as an Electrical/I
and management at DuPo
technical review presentat
training sessions.

I look forward to the chanc
company's industry leader
safety, and excellence are
part of the team and contri

Enclosed is my resume wi
education, work, and othe

Sincerely,

Jane A. McMurrey
Encl.: resume

Patrick H. McMurrey
1108 West 29
Austin, TX 78703

May 25, 2004

Director of Personnel
Automation Associates
7805 Pearl Creek Drive
Austin, Texas 78706

Dear Director of Personnel:

Please accept this letter and the attached resume as my application for the position of Electronics Engineer you currently have open. My extensive experience with secure communications subsystems should prove useful to your enterprise.

As you'll notice in my resume, I have extensive experience in the design and packaging of advanced workstations. With CyMOS, Inc., I have acted as lead in developing programs to calculate and analyze impedance-controlled logic lines and center-of-gravity calculations on CPU chassis.

To my Electrical Engineering degree from the University of Kansas, I am currently adding a PCT degree in workstation hardware and packaging at the University of Texas here in Austin.

I am available for an interview at just about any time that is convenient for you. Contact me at the phone numbers provided on my resume. I look forward to hearing from you.

Sincerely,

Patrick H. McMurrey
Encl.: resume

Figure 10-9 Examples of application letters. The first paragraph of the letter on the right identifies the position being sought and makes one strong statement about the writer's qualifications. (The fancy headings are not a requirement—just a nice, eye-catching touch.)

ward. But if you handle any of these strategies clumsily, problems of tone occur and you run the risk of projecting the wrong image of yourself.

Table 10-2 on the next page lists some tips on how to write an application letter.

Table 10-2   Tips on Writing Application Letters

- Avoid diving headlong into the details of your background and qualifications in the very first paragraph. Create an introductory paragraph that performs the functions mentioned earlier in this chapter.

- Get a specific name or department to which to address the letter; avoid the "To Whom It May Concern" syndrome.

- Individualize the letter for the addressee. Even if you are in a massive job search and are sending out many letters, avoid sounding as though you're a zombie writing form letters.

- Be sure to mention that your resume is enclosed with the letter.

- Use standard business-letter format in the application letter, as shown in the examples in this chapter and as described in detail in Chapter 4. (Remember to punctuate the salutation with a colon, not a comma!)

- Keep the letter to one page. Keep the paragraphs of the letter short: the first paragraph under 5 lines; the body paragraphs under 10 lines.

- Seek a nice, bright, energetic, positive tone. Watch out for the problems with tone discussed in this chapter. (Get someone to read a rough draft of your letter and describe the kind of personality it projects.)

- Write the letter in terms of the prospective employer's needs or interests, and only minimally your own. Discuss yourself according to the prospective employer's needs.

- Use the full application letter (as opposed to the cover letter) unless the job advertisement specifically requests only the resume.

- Avoid negative discussion of previous employers; generally avoid stating reasons why you left previous jobs.

- Unless specifically requested by the prospective employer, avoid discussion of salary, benefits, or other compensation.

- Although it's acceptable to send out high-quality photocopies of the resume, the letter should be freshly typed or printed out. Make the letter appear as though you prepared it especially for the addressee.

- Avoid spelling, grammar, and usage errors and bad writing at all costs!

HOW TO WRITE A FOLLOW-UP LETTER

Write a follow-up letter when you've not heard from a prospective employer after two weeks, after you've had an interview, when you want to acknowledge a refusal of a job offer, and when you must reject or accept a job offer. The most important use of the follow-up letter is for those situations when you are waiting (and waiting) and have had no word from the prospective employer. (See Figure 10-10 for an example.) To write this type of follow-up letter, consider the following guidelines:

- State the reason you are writing the letter—that is, to inquire about the application letter and resume you recently sent.
- Indicate the date you sent the letter and the resume, and specify the position you were inquiring about.
- Suggest that the letter and the resume might have been lost in the mail or routed incorrectly within the recipient's organization.
- Enclose a copy of the original letter and resume, and state in the follow-up letter that you have enclosed them.
- Tactfully encourage the recipient to let you know the status of the position (indicating that your own decisions are dependent upon it).

---

801 East 31st Street #101
Austin, Texas 78701

3 March 1996

Director of Personnel
Automation Associates
7805 Pearl Creek Drive
Austin, Texas 78706

Dear Director of Personnel:

On February 17, I applied for a position as manufacturing engineer with your firm. Not having heard from you in the two weeks since that time, I'm concerned that my letter may have been lost.

Attached is a copy of the original letter and resume that I sent. As you will see, they detail my work experience, my education, and my sincere interest in working for your company.

If you have already made a decision, I would appreciate hearing from you. For the moment, my availability continues. I look forward to discussing the job and my background with you in person.

Sincerely,

Jane A. McMurrey

Jane A. McMurrey
Encl.: Copy of 2-17 letter and resume

Figure 10-10   Follow-up to an application letter. Although the follow-up letter can be used for different reasons, its most important use is to inquire about the fate of an application letter and resume for which you have received no response.

EXERCISES

Talk with several professional engineers about the application letters and resumes they typically see when hiring new engineers. Ask them questions like the following.

1. How do they "read" resumes: line by line from beginning to end? If they skip around and scan, what do they look for? What catches their eye? How important are specific details such as brand names, model numbers, titles of specifications, and dimensions?

2. What can the engineer who is just graduating and getting started in the profession legitimately put in the work experience section of a resume?

3. Should personal information such as hobbies, community activities, or reading interests be kept out of resumes? If not, what purpose do they serve?

4. What are the typical problems that cause a resume to be ignored? How much does the formatting of a resume contribute to their willingness to read a resume carefully? What effect does heavy use of bold, italics, all caps, and different fonts and font sizes have on the way they read a resume?

5. Are applicants asked to send only a resume or only an application letter? Do they expect to see a simple cover letter (as described in this chapter), or do they expect a full application letter highlighting the applicant's relevant and important qualifications?

6. In their view, what is the chief function of the application letter?

7. Does tone ever cause a problem in these letters? Do details and specifics in an application letter matter, or should the letter be general?

BIBLIOGRAPHY

Beatty, Richard H. *The Resume Kit*. New York: Wiley, 1991.

Bostwick, Burdette F. *Resume Writing: A Comprehensive How-to-Do-It Guide*. New York: Wiley, 1985.

Jackson, Tom. "10 Musts for a Powerful Resume." In *CPC Annual*, 30th ed. Bethlehem, PA: College Placement Council, 1986.

Kennedy, Joyce L., and Thomas J. Morrow. *Electronic Resume Revolution*, 2nd ed. New York: Wiley, 1995.

Lewis, Adele. *The Best Resumes for Scientists and Engineers*. New York: Wiley, 1988.

Munger, Roger. "Technical Communicators Beware: The Next Generation of High-Tech Recruiting Methods." *IEEE Transactions on Professional Communication* 45, no. 4 (December 2002).

U.S. Department of Labor. *Job Search Guide: Strategies for Professionals*. Washington, DC: Government Printing Office, 1993.

U.S. Department of Labor. *Tips for Finding the Right Job*. Washington, DC: Government Printing Office, 1992.

U.S. Department of Labor, Bureau of Labor Statistics. "Chemical Engineers." www.bls. gov/oco/ocos029.htm. In *Career Guide to Industries,* www.bls.gov/oco/cg/cdindex.htm.

U.S. Department of Labor, Bureau of Labor Statistics. "Civil Engineers." www.bls.gov/ oco/pdf/ocos030.pdf. In *Career Guide to Industries,* www.bls.gov/oco/cg/cdindex.htm.

U.S. Department of Labor, Bureau of Labor Statistics. "Electrical and Electronics Engineers." www.bls.gov/oco/pdf/ocos031.pdf. In *Career Guide to Industries,* www.bls.gov/oco/cg/ cdindex.htm.

U.S. Department of Labor, Bureau of Labor Statistics. "Mechanical Engineers." www.bls.gov/ oco/pdf/ocos033.pdf. In *Career Guide to Industries,* www.bls.gov/oco/cg/cdindex.htm.

U.S. Department of Labor, Bureau of Labor Statistics. "Petroleum Engineers." www.bls.gov/ oco/pdf/ocos037.pdf. In *Career Guide to Industries,* www.bls.gov/oco/cg/cdindex.htm.

Weddle, Peter D. *Electronic Resumes for the New Job Market.* Manassas Park, VA: Impact, 1995.

11

DOCUMENTATION AND ETHICS IN ENGINEERING WRITING

The days when an engineer's only ethical commitment was loyalty to his or her employer have long passed. The expansiveness of technology is such that now, more than ever, society is holding engineering professions accountable for decisions that affect a full range of daily life activities...and for engineers, the implications are inescapable. Handling ethical dilemmas and making ethical decisions are very important elements of being a professional.

Murdough Center for Engineering Professionalism, Texas Tech University, www.niee.org

If it is not right do not do it; if it is not true do not say it.

Marcus Aurelius

A common problem among students in high school and college is plagiarism, but unfortunately this kind of dishonesty is not limited to youth or the academic world. If as an engineer you knowingly or unknowingly "borrow" the language, ideas, thoughts, or graphics of others, representing them as your own original work by failing to acknowledge your sources, you are plagiarizing—a very serious offense. You might even be infringing on someone's copyright, and thus could open yourself up to the possibility of lawsuits.

Plagiarism is frequently the result of ignorance or carelessness rather than dishonesty. Some writers and researchers simply get lazy: It's easier to replicate another's ideas or style than to think about what you have read and then put it in your own words and reference it. However, when you do research, *all information* that you obtain from journals, books, interviews, the Internet, and so on *must* be fully documented—that is, accompanied by references to the sources where you obtained the information. Only your own ideas and opinions, plus common knowledge, need not be referenced.

COMMON KNOWLEDGE

The only exception to the rule of acknowledging your sources is when you cite common knowledge. But what is "common knowledge"? What may be common knowledge to some may not be common knowledge to others. Is anything in the engineering world "common knowledge"? Think of a theory you learned in engineering school: You can find it in practically every standard textbook on the subject, and it is not documented when it is discussed in those textbooks. That's common knowledge. But think of a current theory put forth by an engineer who is not well known in his or her field. That's not common knowledge, and if you used the theory in a report, you would have to document your source for it. The difference then comes down to your familiarity with your field, and whether you can distinguish what is common knowledge to your audience from information that is not.

CITING INFORMATION

How you cite your sources will depend on the documentation system you use (we provide a standard system in this chapter), but again, your reader must *always* know which ideas and judgments are yours and which you obtained by consulting the work of others. This point cannot be stressed too often. If you don't provide source information by means of thorough and reliable documentation, the credibility of your work will collapse like a house of cards. In summary, you document your information borrowings in order to

- Protect the originator, the author of the information, so that she or he will get the credit and acknowledgment for having developed it
- Protect yourself from accusations of plagiarism—of stealing other people's hard-won discoveries
- Demonstrate to readers that you have done your homework and are aware of the latest developments in the particular field
- Enable readers to track down the information so that they can read it for themselves

Note With the evolution of the Web, plagiarism is more prevalent and harder to spot because of the limitless amount of information now available via our keyboards. You can access libraries, reports, journals, and graphics within seconds. Any of this information that helps you in your research must be documented, including images or graphics from the Web. To counteract the rampant plagiarism that now takes place in colleges and universities, professors are currently using programs that can search the Web and find whether specific pages, paragraphs, and even sentences are stolen (i.e., not documented) from any area of the Web.

A SYSTEM FOR DOCUMENTING YOUR SOURCES

The following pages give examples of how to document your sources in a research paper. This information is based on the system generally used by the Institute of Electrical and Electronics Engineers (IEEE), but it is similar to many of the systems used by other engineering organizations and industries. First we give procedures for inserting reference numbers in your text. Next are guidelines on how to format a references page. Finally, we provide a list showing how to format the numerous and varied sources you might use in your research when listing them on a references page.

PROCEDURES FOR DOCUMENTING SOURCES

1. In the body of your text, refer to the source of your information by inserting consecutive numbers in brackets, beginning with 1, at the end of each segment of cited information—like this [1]. This tells the reader the borrowed information came from source 1 on your references page. Reference numbers can also be inserted within a sentence like this [2], without changing the sentence's punctuation. You may also cite your reference in your text thus: *According to the 2002 U.S. Census Bureau [3], we see that. . . .* Notice that a space always precedes the bracketed number and that the punctuation comes after the second bracket.

2. Unless you are referring to a complete book or article, you will need to identify the page number(s) of your source of information. Indicate exact page numbers of a source within your brackets after a comma [4, pp. 3–6], or by a simple rhetorical device in your text, such as *However, on page 79 of [5] the author seems to contradict herself when she states. . . .* If you must refer to more than one source in the same reference—because you have combined information from more than one source in a paragraph, for example—use semicolons for separation: [6, p. 46; 7, pp. 29–31; 9, pp. 8, 12].

3. References at the end of quotation marks are punctuated with the period after the reference, "like this" [8, p. 23]. Once you have numbered a source, use the same number for all subsequent references to that source throughout your work.

Figure 11-1 shows a section from a research paper that is documented following these specifications.

8

6.0 THE FUTURE OF HEVS

Knowing exactly what the future holds for HEVs is impossible. However, using what we know to be true today, we can generally extrapolate to a reasonable degree what tomorrow might bring.

6.1 Options

With technology comes options, and hybrid technology is no different. There are many different ways in which a hybrid can be configured, and since each has its own advantages, many different options will most likely be offered to the consumer. "Rather than having only one propulsion system choice when buying a future vehicle, it may be possible to select the propulsion system in the same way that one selects a 4 cylinder engine or a V 8" [10, p. 43]. One could choose from a conventional gasoline, battery only, or any number of configurations of an energy storage device and a hybrid power unit (HPU) [9, pp. 98–99].

6.2 Fuel cells

Though today's HEVs have a conventional gasoline or diesel engine combined with an electric motor, in the next five years we will most likely see the arrival of the fuel cell in hybrid vehicles [13, p. 11]. Much work—and money—is going into improving on this technology.

6.2.1 Brief overview of the fuel cell. Fuel cells generate electricity through an electrochemical reaction that combines hydrogen with air. Many different fuels can be used, but methanol is often the fuel of choice, with which the fuel cell's only emission is water vapor, making it the cleanest alternative available [1].

6.2.2 Current limitations of fuel cells. Unfortunately, fuel cells need further development in order for them to be feasible in personal automobiles. First of all, as with all new technology, the fuel cell is expensive. It will take some deflation of cost before it can match the cost of a conventional gasoline engine, and thus penetrate the market [16, pp. 14–16]. In addition, the fuel cell has not been a viable option due to its large size. However, great strides have been made in this area in the past few years, and "officials at DaimlerChrysler have pledged to have a viable, commercial fuel cell vehicle available in 2004" [16, p. 17].

(continues)

Figure 11-1   A section from a well-documented research paper.

In order to reform fuel (change it into its useful form so it can react to create energy), the system has to be heated to a certain temperature in order for the reaction to occur [13, p. 8]. Thus, long start-up times are also holding fuel cells back from use in HEVs, yet although there are still considerable strides to be taken in fuel cell technology, these cells will definitely serve as a viable option for HEVs in the near future [1].

6.3 Future models

Only two car companies have HEVs on the market today, but in the next few years almost all car companies are likely to follow suit [9]. As they flood the market, prices will drop, and the HEV will be cost comparable to a conventional vehicle. Below are some HEV models that might be emerging in the next few years.

6.3.1 Ford P2000 LSR. One model to be introduced shortly is the Ford P2000 LSR, which was delivered by the Ford Motor Company to the U.S. Energy Department in October, 1999. The P2000 LSR will be a hybrid diesel-electric vehicle with "the passenger room, trunk space and driving acceleration of a Taurus" [17]. Ford has also designed the Ford Prodigy, a concept, diesel-electric hybrid family sedan that will get 80 miles to the gallon [18, p. 3].

Figure 11-1 (*continued*)   A section from a well-documented research paper.

FORMAT OF THE REFERENCES PAGE

- Do *not* list your sources by alphabetical order of authors' last names, but in numerical order according to when they are first cited in the text.
- Note that in the IEEE system, only the initials of authors—*not* their full first names—are given. Also, the titles of journal articles are given in sentence form rather than title form.
- Single space individual references, with a second or third line aligned with the first. Double space between separate references.
- Use a common abbreviation for a journal title if there is one (e.g., *IEEE Electron Device Lett.*) Otherwise, give the full name of the journal.
- End each entry with a period.
- Even if you have referred to the same source more than once in your paper, list that source only once on your references page.

Figure 11-2 provides an example of a references page using these guidelines.

SAMPLE REFERENCES

Following are examples of items that would be listed on a references page. They illustrate most of the kinds of references you will likely have to cite. If you come across some sources of information that you have no model for citing, simply be

---

REFERENCES

[1] C. H. Roth, *Fundamentals of Logic Design*, 4th ed. St. Paul: West Publishing Company, 1992.

[2] R. Schneiderman, *Future Talk: The Changing Wireless Game*. New York: IEEE Press, 1997.

[3] N. Hart, "Mobile satellite system design." In M. J. Miller, ed., *Satellite Communications: Mobile and Fixed Services*, pp. 103–143. Boston: Kluwer Academic Publishers, 1993.

[4] D. Pearl, "FAA clears civilian airlines to use military satellite signals in navigation," *The Wall Street Journal*, p. A18, February 18, 1994.

[5] *GPS NAVSTAR User's Overview*. Los Angeles: ARINC Research Corporation, 1991.

[6] Personal interview with Dr. Francis Bostick, ECE Department, The University of Texas at Austin, November 18, 2001.

[7] F. Vizard, "In trouble? Call Ford," *Popular Mechanics*, vol. 172, p. 32, July 1995.

[8] C. Hilary and D. Mor, "The power infrastructure," http://www.cs.dartmouth.edu/2K/power-CM/ Accessed April 2, 2001.

[9] S. J. Childe, R. S. Mall, and J. Benett, "Frameworks for understanding business process re-engineering," *Int. J. Oper. Prod. Manag.*, vol. 14, no. 12, pp. 22–34, 1994.

[10] Email from Mark A. Carpenter, A98-b2 project manager, AMD, Austin, Texas, March 8, 2003.

Figure 11-2   An example of a brief references page using IEEE style.

guided by the needs of your audience; that is, provide enough information to allow your readers to go to that source if they want to.

Book

[1] B. P. Lathi, *Linear Systems and Signals*. London: Oxford University Press, 2001.

Book, Multiple Authors

[2] S. Horner, T. Zimmerman, S. Dragga, *Technical Marketing Communication*. New York: Longman, 2002.

New Edition of a Book

[3] C. Conrad and M. S. Poole, *Strategic Organizational Communication,* 5th ed. New York: Harcourt Press, 2002.

Journal Article

[4] R. F. Boehm, "Heat engineering," *Developments in the Design of Thermal Systems*, vol. 16, no. 6, pp. 190–206, June 1997.

Article in an Anthology

[5] G. J. Broadhead, "Style in technical and scientific writing." In M. G. Moran and D. Journet, eds. *Research in Technical Communication: A Bibliographic Sourcebook*, pp. 379–401. Westport, CT: Greenwood Press, 1985.

Translation

[6] M. M. Botvinnik, *Computers in Chess: Solving Inexact Search Problems*. Translated by A. Brown. Berlin: Springer-Verlag, 1984.

Personal Interview/Communication

[7] Interview [or Personal communication] with Prof. David Beer, ECE Department, The University of Texas at Austin, January 10, 2003. [Date omitted if unknown.]

Handbook or Data Book, No Author

[8] *Handbook of Accelerator Physics and Engineering*. Singapore: World Scientific Institute, 1999.

[9] *Engineering Ceramics Data Book*. Engineering Materials Series. New York, 1998.

[10] *User's Guide: Microsoft Word*. Vers. 6.0. Microsoft, 1995.

[11] HMC224Ms8GaAsMMIC T/R Switch Data Sheet, Hittite Microwave Corporation, 2001.

Encyclopedia Entry

No author given:

[12] "Frequency," Encyclopedia Britannica, 2001 ed.

Author(s) given:

[13] D. G. Paxon, D. S. Wood, and W. C. Malden, "Equity," in *The Blackwell Encyclopedia of Finance*, F. Carter, ed. Malden, MA: Blackwell Publishing, 1999.

Online:

[14] "Thermodynamics," *The New Online Britannica*, April 2002. http://search.eb.com/

Course Notes

[15] A. S. Erickson, *Lab Notes for EE464K, Senior Projects*, The University of Texas at Austin, Spring semester, 2003.

Dissertation or Thesis

[16] G. Davis, "Adaptive nonlinear approximations," Ph.D. dissertation, New York University, New York, Sept. 1994. [Add if applicable: University Microfilms, Inc., University of Michigan, Ann Arbor, Michigan.]

Proceedings Paper

[17] N. Coppola, "Computer-based training for chemists: Designing decision-making tools for green chemistry," in *Proceedings of the International Professional Communication Conference*, pp. 77–83, Portland, OR, Sept. 17–20, 2002.

Patent

[18] M. L. Chirinos, U.S. Patent 5 670 087, 2001. [Title of patent may be included.]
[19] M. Postol, "Method of lattice quantification which minimizes storage requirements and computational complexity," U.S. Patent 6 085 340, July 4, 2000.

Newspaper Article

[20] "Virus overwhelms global Internet systems," *The New York Times,* vol. 116, pp. A3, A8, January 27, 2003.

Government Publication

[21] *Basic Facts about Patents.* Washington D.C.: Government Printing Office, 2002.

Technical Report

[22] R. Cox and J. S. Turner, "Project Zeus: design of a broadband network and its application on a university campus," Washington Univ., Dept. of Comp. Sci., Technical Report WUCS-91-45, July 30, 1991.
[23] "TDDB results for 0.18 μm," Taiwan Semiconductor Manufacturing Co., Hsinchu, Taiwan, R.O.C., 2001.

Letter or Email

[24] Letter [or Email] from A. R. Hasan, Project Manager, Oracle, Boston, Massachusetts, Jan. 5, 2003.

Software

[25] J. McAfee, *Virus Scan Version 6.0.* Computer software. Networks Associates Technology, Inc. IBM-PC, 2001.

Database/Online

[26] R. Berdan and M. Garcia, *Discourse-Sensitive Measurement of Language Development in Bilingual Children* (Los Alamitos, CA: National Center for Bilingual Research, 1982) (ERIC ED 234 636).
[27] J. Ozer, "External solutions for your expanding video library," *PC Magazine*, Jan. 27, 2003, v22, n10 p. 247(7) in Academic Index (database on UTCAT PLUS system).

World Wide Web

[28] "AT&T enters Indiana residential local phone market," http://www.att.com Accessed Jan. 26, 2003.
[29] "Nokia introduces the world's first handset for WCDMA and GSM networks," http://press.nokia.com/pr2002_3.html Accessed Jan. 27, 2003.
[30] B. L. Evans, "Brian Evans' home page," http://www.ece.utexas.edu/~bevans/ Accessed Feb. 12, 2003.

Slides and Films

[31] L. J. Mihalyi, *Landscapes of Zambai, Central Africa.* Santa Barbara, CA: Visual Education, 1975. (slides)

[32] *An Incident in Tiananmen Square*, 16 mm, 25 min., Gate of Heaven Films, San Francisco, CA, 1990. (film)

Videocassette/DVD

[33] *Behind the Lines*. 96 min. Artisan Entertainment, 1997. (videocassette)
[34] F. W. McMaster, *Matrix Algebra for Electronic Circuit Analysis*. Cottage Publishing, Flower Station, Ontario, Canada. (video instruction tape) No date.
[35] The Standard Deviants: Physics, Part 2. Cerebellum Corp., 1999. (DVD)

ENGINEERING ETHICS

Technology has a pervasive and profound effect on the contemporary world, and engineering plays a central role in all aspects of the development of technology. Because of this it is vital that there be an understanding of the ethical implications of engineers' work. Engineers must be aware of their social responsibilities and equip themselves to reflect critically on the moral dilemmas they will confront.

Mike Martin and Roland Schinzinger, *Ethics in Engineering*, 2nd ed. (New York: McGraw-Hill, 1989), p. xv

It's hard to live very long without making numerous decisions affecting both your own well-being and that of others. Engineers are no different from anyone else: During your career you will have to make countless choices among various courses of action. Thus, you should be familiar with some of the factors involved in making ethical choices. You should also have an idea of the kinds of situations that will require you to make responsible decisions as an engineering writer. After reading this section, you will be aware of pitfalls to avoid as a writer and resources to help you avoid them.

Wherever you find technology, you're going to find ethical and moral concerns. Manufacturing and selling an automobile or any other piece of equipment when it is known to be unsafe is an ethical, as well as a legal, matter. Where to dump hazardous waste raises considerable moral questions, as do the issues, for example, of building with asbestos or locating high-power transmission lines. Accurate record keeping, the ethical use of software, or professional consulting outside your regular job are also examples of the varied activities or situations where you as an engineer might find yourself having to make moral choices. While working with other people, you may also at times be confronted with issues of dishonesty, discrimination, harassment, and alcohol or drug abuse—all situations calling for sound ethical decisions.

This section focuses on concerns you must be aware of as an engineering writer and researcher. Some of these "concerns" are actually illegal practices engineers sometimes commit either knowingly or unknowingly. In some instances they have paid heavy prices for their actions, such as lawsuits, job loss, or at least a diminished reputation. Let's look at five major areas where problems can arise for engineers who research and produce information to be shared with others in writing.

COPYRIGHT INFRINGEMENT

Many people know very little about copyrights and think that because an image or article is available in print or on the Internet, they have a right to copy and use it. This is simply not so. If you come up with original ideas or inventions as a result of your own research, these ideas or items are protected under law if you copyright them. A copyright is basically the legal right you (or your company) are granted to enjoy complete possession and profits from your work for a certain time. To obtain a copyright is a fairly simple procedure: You file a copyright form, pay a filing fee, and provide the Copyright Office with one or more copies of the work to be copyrighted. Once you have copyrighted your work, it cannot be used or distributed without your permission (with a few minor exceptions). If someone does so, they have infringed on your copyright and you may be able to sue them.

We have perhaps oversimplified this brief discussion of copyrights to get a point across. (You can find complete information about copyrights at the U.S. Copyright Office website at www.loc.gov/copyright/.) For the engineering researcher and writer, the important point is *always* to be aware of what is someone else's intellectual property and to never use it or cite it in any way without permission or acknowledgment. The one exception to this is that you don't need permission to quote or paraphrase a small amount of copyrighted work for educational purposes, as long as you give credit to the source and gain no financial profit from your use of the work.

TAMPERING WITH RESULTS

Engineers often have to write up the results of their research and experimentation. What if the numbers don't quite come out the way they were supposed to? Perhaps you are part of a team working on a suspension bridge and run into a small problem toward the end of construction. The team decides the problem can be overlooked if they change a few measurements to meet given requirements. In your final report, would you carefully change a few numbers so that things "come out right"?

Engineers must deal with this sort of moral issue sometimes. An ethical engineer wouldn't change any results and would work until the problems had been solved and everything was accurate. Sometimes it might seem that a few changed details won't hurt, but tampering with results is a very serious issue in the engineering field and is a choice that sooner or later can come back to haunt you.

Another form of tampering with results is found in concocting data. Here a writer makes up information or results with no backing or truth behind them—they are fictitious. Unethical engineers (and others!) have been known to insert concocted data in reports to show progress or results that are nonexistent, often in order to get further funding or to hide a lack of real effort. Again, time has a way of uncovering such actions.

Withholding Adverse Information

Plenty of engineering evidence shows that withholding adverse information can lead to problems, accidents, and even deaths. Ford Pintos and Firestone tires immediately come to mind in this context. Our point here is that if any kind of damage results because you withhold information about a flawed design, a dangerous product, or a means to avoid harm, it is your or your company's responsibility. You can certainly be held liable for your inaction. No ethical engineer should keep silent or fail to include in a written report anything concerning a product or process that might result in a user's financial loss, physical harm, or death.

Withholding adverse information can also occur in job applications and resumes. There is often a temptation to omit less admirable events in one's past in these documents, just as there might be an urge to concoct data for them. Of course, you don't have to put *everything* in your resume, and it is perfectly justifiable to focus on your strong points when writing job-application documents. However, as several well-publicized cases in recent years have shown, many companies and institutions maintain strict policies that enable them to fire workers who falsify resumes in any way. Thus, you are helping yourself when you write these—and all—documents in an ethical manner.

Writing Unclear Instructions

As an engineer you may well be involved in writing instructions, procedures, manuals, or user guides at some point in your career. No matter what you call them, directives telling someone how to do something need to be written in a detailed and precise manner. There is no room for error or ambiguity. Unclear instructions on operating an aircraft, space shuttle, or nuclear reactor might result in chaos or death, and ambiguity about assembling or operating everyday products—such as computers, cameras, pumps, filters, or telescopes—will cause frustration and anger, plus a diminished respect for the product and the company that produced it.

Examples of unclear instructions and their consequences abound. Many of us know that sinking feeling when we see the words "Assembly Required." The problem is rarely one of malicious intent but rather of poor planning on the writer's part, careless or hasty writing and editing, or a failure to put oneself firmly in the head of the reader or user. By studying the section on instructions in Chapter 5 of this book, you will gain a good background on writing effective instructions and procedures and should be able to avoid problems. The more skilled you become at producing watertight directions, the less likely you will be to frustrate your readers, anger them, endanger them, or be sued by them.

OMITTING SAFETY WARNINGS

Engineers should constantly be concerned with the safety of their customers and of anyone else their products and designs might affect. This means you must write clear and conspicuous safety warnings into the description of any design, procedure, or product that requires them, because you are always responsible for providing information that ensures the consumer's safety. Failure to provide adequate safety warnings in your writing can, depending on what you are describing, lead to mishaps, loss, disaster, serious physical harm, or even death. To avoid this, you should always take great care to provide clear warnings whenever necessary in your writing—and you should ensure they are visually prominent and accessible to your reader.

Although one of the most basic canons of engineering is to put the well-being of the public first, there may be times when you feel subtle pressure from a company or manager to not stress safety problems when describing a product. At this point you are facing an ethical issue similar to that of withholding adverse information. As with ethical dilemmas in other situations, a wrong choice may come back and bite you with a vengeance later on.

TOOLS FOR ETHICAL DECISION MAKING

There is no need to despair when faced with any of the problems just described, because there are tools you can use to justify doing the right thing. Some of the most powerful are the codes of ethics published by professional engineering associations and by some of the larger engineering firms. You can find many of them online by entering "Code of ethics of engineers" in any good search engine. With these in hand, you can refer to specific tenets and guidelines that will back you up and verify your decisions to hold out for strictly ethical writing (and other activities) as an engineer.

Two such codes are shown in Figures 11-3 and 11-4. You can always use them as support if you find you have to defend your decisions against any implied or real pressure. Following the codes of ethics is a suggested checklist for ethical decision making (Table 11-1) that you might also find personally useful when uncertain about what choice to make or what plan of action to take.

MAKING UP YOUR MIND

It would be nice if all choices of action were simply between right and wrong, good and bad, or ethical and unethical. Although there is no question about the writing

Accreditation Board for Engineering and Technology

CODE OF ETHICS OF ENGINEERS

THE FUNDAMENTAL PRINCIPLES

Engineers uphold and advance the integrity, honor and dignity of the engineering profession by

I. using their knowledge and skill for the enhancement of human welfare;

II. being honest and impartial, and serving with fidelity the public, their employers and clients;

III. striving to increase the competence and prestige of the engineering profession; and

IV. supporting the professional and technical societies of their disciplines.

THE FUNDAMENTAL CANONS

1. Engineers shall hold paramount the safety, health and welfare of the public in the performance of their professional duties.

2. Engineers shall perform services only in the areas of their competence.

3. Engineers shall issue public statements only in an objective and truthful manner.

4. Engineers shall act in professional matters for each employer or client as faithful agents or trustees, and shall avoid conflicts of interest.

5. Engineers shall build their professional reputation on the merit of their services and shall not compete unfairly with others.

6. Engineers shall act in such a manner as to uphold and enhance the honor, integrity and dignity of the profession.

7. Engineers shall continue their professional development throughout their careers and shall provide opportunities for the professional development of those engineers under their supervision.

ABET

345 East 47th St., New York, NY 10017

1987

Figure 11-3 A typical code of ethics for the engineering profession. You may use documents like this to support your position when faced with an ethical choice of action.

topics discussed earlier, some problems are not so easily delineated. Sometimes you will be faced with complex professional situations where you have to consider the issues involved from more than one angle. You will then need to evaluate the problem in relation to other moral values, constraints, and concerns, including your own, your company's, and those of the society in which you live. You may also need to consult with others.

THE INSTITUTE OF ELECTRICAL AND ELECTRONICS ENGINEERS, INC.

Code of Ethics

We, the members of the IEEE, in recognition of the importance of our technologies in affecting the quality of life throughout the world, and in accepting a personal obligation to our profession, its members and the communities we serve, do hereby commit ourselves to the highest ethical and professional conduct and agree:

1. to accept responsibility in making engineering decisions consistent with the safety, health, and welfare of the public, and to disclose promptly factors that might endanger the public or the environment;

2. to avoid real or perceived conflicts of interest whenever possible, and to disclose them to affected parties when they do exist;

3. to be honest and realistic in stating claims or estimates based on available data;

4. to reject bribery in all its forms;

5. to improve the understanding of technology, its appropriate application, and potential consequences;

6. to maintain and improve our technical competence and to undertake technological tasks for others only if qualified by training or experience, or after full disclosure of pertinent limitations;

7. to seek, accept, and offer honest criticism of technical work, to acknowledge and correct errors, and to credit properly the contributions of others;

8. to treat fairly all persons regardless of such factors as race, religion, gender, disability, age, or national origin;

9. to avoid injuring others, their property, reputation, or employment by false or malicious action;

10. to assist colleagues and co-workers in their professional development and to support them in following this code of ethics.

Approved by the IEEE Board of Directors, August, 1990

Figure 11-4 The ten ethical guidelines used by the IEEE. These also could be used to substantiate an ethical position you feel you must take.

In other words, when faced with professional dilemmas in your engineering career, whether or not written work is involved, you must analyze the situation carefully and responsibly. Like a good engineer, you should think long and hard before making a decision. To help do this, you may wish to consider the set of questions in Table 11-1.

Table 11-1 Checklist for Ethical Decision Making

- ☐ What caused this dilemma in the first place?
- ☐ Have I clearly defined the dilemma and its possible options?
- ☐ Should others be involved in any final decision?
- ☐ What are the immediate and long-term results of each option likely to be?
- ☐ Could any option injure anyone (a) physically (b) emotionally (c) professionally?
- ☐ Are all my options legal?
- ☐ To what extent does each option follow the "golden rule"?
- ☐ Will my decision be one I would willingly share with my
  - ☐ management
  - ☐ colleagues
  - ☐ family
  - ☐ lawyer
  - ☐ local news media
  - ☐ religious leader?
- ☐ Whatever option I choose, could there ever be exceptions to it?

A FINAL WORD

Our personal ethics are often determined by our personal philosophy of human existence. Do you feel that the prime goal of human endeavor is simply to survive, or to achieve unlimited pleasure? To gain unlimited possessions, or to live in harmony with nature? To live happily with one another, or according to the dictates of a divine power? You are really the only one who can validly answer these questions for yourself, although others might have told you how *they* feel you *should* think or act.

Whatever your personal outlook, it's worth remembering that the study of ethics will not necessarily make you a "better" person, but it will make you a more knowledgeable person when you come face to face with difficult professional decisions. We hope that this chapter has given you some insight and tools that will allow you to be an ethical researcher, writer, and engineer.

EXERCISES

1. Go to a good Web search engine and enter "Codes of ethics for engineers." Either read some of the professional codes on-screen or print them out. Study them carefully. How are they similar and how are they different (if at all)? Think of situations that might arise in your career where you would be glad to have such codes to support you in your actions.

2. Access the Murdough Center for Engineering Professionalism's website at www.niee.org. Explore the various topics that are included in this site, particularly the case studies of actual engineering problems. What do you learn from them? Why do you think such a center as the Murdough one is so important to the profession?

3. Ask any engineers if they have had to make ethical decisions in their career. What was the nature of the dilemma? How did they make their decision? Were there any repercussions? Were they happy with their decision?

4. Research the disasters of the *Challenger* and *Columbia* space shuttles. What ethical questions did these disasters raise, both from a human and an engineering perspective? You may also wish to investigate other well-known tragedies in the automobile, nuclear, shipping, construction, or other industries. What do you learn from them regarding ethics and engineering decisions?

BIBLIOGRAPHY

Davis, Michael. "Thinking Like an Engineer: The Place of a Code of Ethics in the Practice of a Profession." *Philosophy & Public Affairs* 20, no. 2 (Spring 1991), 150–167.

Jones, Dan, and Karen Lane. *Technical Communication: Strategies for College and the Workplace*. New York: Longman, 2002. See particularly "Appendix C: Documentation Styles."

Landis, Raymond B. *Studying Engineering: A Road Map to a Rewarding Career*, 2nd ed. Los Angeles: Discovery Press, 2000.

Martin, Mike W., and Roland Schinzinger. *Ethics in Engineering*, 3rd ed. New York: McGraw-Hill, 1996.

Radford, Marie, Susan Barnes, and Linda Barr. *Web Research: Selecting, Evaluating, and Citing*. Boston: Allyn and Bacon, 2002.

INDEX